The Water-Food-Energy Nexus in the Mekong Region

Alexander Smajgl • John Ward
Editors

The Water-Food-Energy Nexus in the Mekong Region

Assessing Development Strategies Considering Cross-Sectoral and Transboundary Impacts

Springer

Editors
Alexander Smajgl
CSIRO Ecosystem Sciences
Townsville, QLD, Australia

John Ward
CSIRO Ecosystem Sciences
Townsville, QLD, Australia

ISBN 978-1-4614-6119-7 ISBN 978-1-4614-6120-3 (eBook)
DOI 10.1007/978-1-4614-6120-3
Springer New York Heidelberg Dordrecht London

Library of Congress Control Number: 2012953575

Printed on acid-free paper

Springer is part of Springer Science+Business Media (www.springer.com)

Acknowledgments

The AusAID CSIRO Research for Development Alliance provided the funding for the Exploring Mekong Region Futures research project. This volume documents the deliberations and reviews of an expert panel conducted as an important aspect of the project's overall research methodology. We are grateful for the valuable time and resources the experts enthusiastically contributed during the expert panel assessment. We especially thank the workshop participants for their efforts in cohering their individual Mekong region experiences and judgements into the collective wisdom articulated in this volume. The staff of the AusAID mission in Vientiane provided crucial and sustained support, particularly the enthusiasm, impartiality and sagacity of John Dore. We also acknowledge the leadership of the CSIRO Climate Adaptation Flagship and the Division of Ecosystem Sciences for their foresight and commitment. Finally, we acknowledge the substantial manuscript improvements due to the critical appraisal, corrections and insights suggested by several reviewers.

Contents

Glossary and Abbreviations

ADB	Asian development bank
Alumina	Aluminium oxide refined from bauxite ore feedstock for the smelting of aluminium metal
Aluminium refining	Bauxite ore is finely crushed and dissolved in a solution of sodium hydroxide (caustic soda, or lye) under high temperature and pressure. Insoluble iron oxide, titanium, sodium, silica and other oxides are filtered out as sludge called 'red mud'. The solution is then clarified and sent to a precipitation tank where a small amount of aluminium hydroxide is added as a 'seed' that facilitates the crystallization of aluminium hydroxide and sodium hydroxide. The crystals are then washed, vacuum dewatered and sent to a rotating kiln. The result is a fine white powder called alumina (aluminium oxide).
Aluminium smelting	The strong bonds between aluminium and oxygen in alumina makes its refining into aluminium possible only by using enormous amounts of energy, more than that required in the production of any other metal, or in fact in any other industrial process. Primary aluminium processing is the most polluting phase of the aluminium production chain, resulting in air emissions and solid wastes.
ARWR	Annual renewable water resource
BAU	Business-as-usual (referring to incremental changes that occur in a system)
Bauxite	Aluminium ore, found principally in tropical and sub-tropical areas
Bayer process	Procedure used to refine bauxite ore into alumina
BDP	Basin development plan
BS	Baseline scenario
DF	Definite future

Dutch disease	Term for the negative relationship between resource abundancy and GDP growth also often referred to as resource curse. The underlying mechanism assumes that increasing income from resource exports can lead to a decline in the relative prices (terms-of trade) between trade and non-trade sectors, affecting investments and therefore growth.
EE	Energy efficiency
EMRF	Exploring Mekong Region futures
ERWR	External renewable water resource
Final energy	Includes primary energy minus inputs for (electricity, heat, refineries and other energy) plus (energy value of electricity). *Examples:* Petroleum, electricity, fuel wood.
GMS	Greater Mekong subregion
GOL	Government of Lao PDR
GWh	Gigawatt-hour (equals one million kilowatt hours)
Industrial metals	gypsum, limestone, silica sand and kaolin
IPCC	Intergovernmental panel on climate change
IRWR	Internal renewable water resource
LECS	Lao expenditure and consumption survey
LMB	Lower Mekong basin
LMD	Lower Mekong mainstream dam
MCM	Million cubic metres
MD	Mainstream dam
MRB	Mekong river basin
MRC	Mekong river commission
MWh	Megawatt-hour (equals 1,000 KWh)
MWyr	Megawatt-year (equals 8.76 GWh)
N	Nitrogen
NTFP	Non-timber forest product
P	Phosphorus
PJ	Petajoule (10^{15} Joule, equivalent to 277.78 GWh, and 23.88 kiloton oil equivalent)
POE	Panel of experts
PRC	People's Republic of China
Primary aluminium	Aluminium ingots produced from bauxite and other aluminium ores via a smelting process
Primary energy	Energy embodied in natural resources prior to human conversion or transformation. *Examples:* Coal, crude oil, uranium, solar, wind.
RE	Renewable energy
Red mud	Bauxite residue from refining to alumina via the Bayer process. For every ton of alumina produced, between 2 and 3 t of bauxite ore must be processed. The waste remaining after the process is disposed of is red mud.

SEA	Strategic environmental assessment
TMD	Thai mainstream dam
TWh	Terawatt-hour (1,000 GWh)
UMB	Upper Mekong basin
UMD	Upper mainstream dam
Useful energy	Energy that provides end users with energy services such as cooking, illumination, space conditioning, transport and industrial heating
VN	Vietnam
WMD	Water management device

Chapter 1
Mekong Region Connectivity

Alex Smajgl and John Ward

1 Rationale

This book provides a cross-sectoral, multi-scale assessment of development-directed investments in the wider Mekong Region. The wider Mekong Region includes Lao PDR, Cambodia, Thailand, Vietnam, Myanmar and the Chinese Province of Yunnan, see Fig. 1.1. Evidence highlights a limited set of critical dynamics, including human migration, natural resource flows and increasing levels of private and State financial investments generate a high level of connectivity between these countries (Contreras 2007; Dore 2003; Harima et al. 2003; Theeravit 2003).

For example investors from China, Thailand and Vietnam increasingly replace traditional donor organisations in order to source natural resources or manufacturing capacity in neighbouring countries (Middleton et al. 2009; Molle et al. 2009).

High levels of connectivity increase complexity, biasing the reliability of predicted outcomes and increasing the potential for unforeseen consequences of national decisions (Cechich et al. 2003; De Landa 2006; Sawyer 2005). The effects of large scale investments in weakly connected regions are generally constrained to locales proximate to the initial investment area within a particular country and tend to be limited to the investing sector. In contrast high connectivity implies that investment factors (or drivers) interact, transmitting the effects of substantial changes from one part of a region to another and to other sectors.

However, the degree of connectivity itself is an emerging phenomenon. In situations of increasing scarcity and constraints on the availability of required production resources, incentives to source additional resources from elsewhere also increase. Such resources might include labour, electricity, water, agricultural produce, or minerals. The emerging links between countries can unfold in financial investments, migration or the flow of resources. As these links intensify the regional connectivity increases and a highly connected region can emerge, as experienced by the Mekong Region.

Thus, connectivity is an emerging phenomenon shaped by these three conditions. If the nature of one of these enabling conditions changes, the shape of the highly

A. Smajgl and J. Ward, *The Water-Food-Energy Nexus in the Mekong Region: Assessing Development Strategies Considering Cross-Sectoral and Transboundary Impacts*, DOI 10.1007/978-1-4614-6120-3_1, © Springer Science+Business Media New York 2013

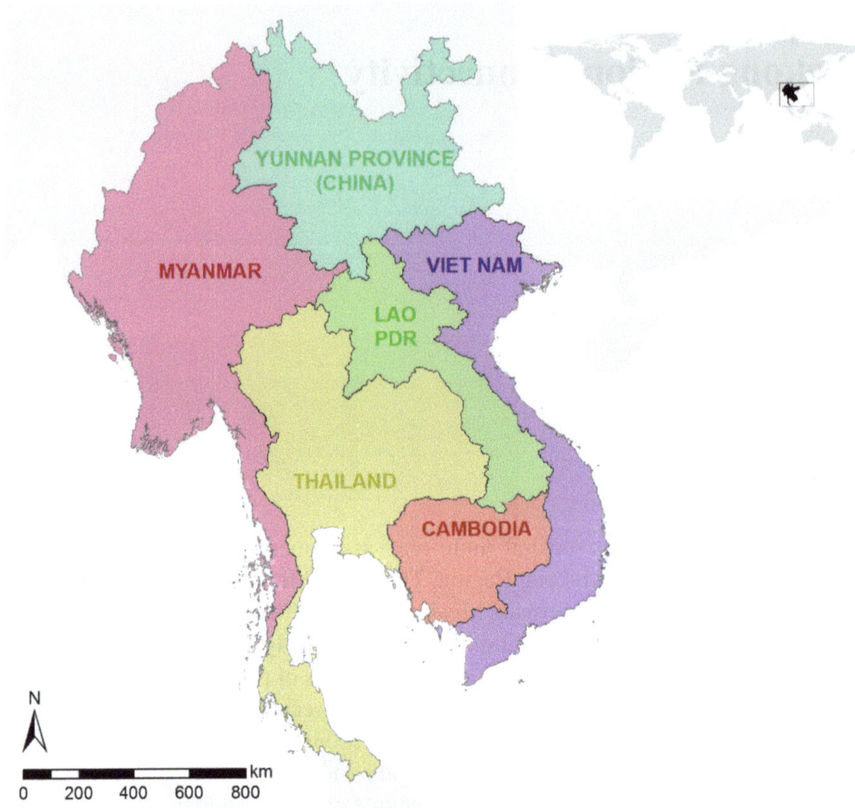

Fig. 1.1 Map of the wider Mekong Region

connected region is also likely to change. For instance, if the biophysical reality changes and required resources are not available from traditional locales but become available in another area outside of the region, the new area is likely to become part of the highly connected region, irrespective of national borders.

Three enabling conditions that are central to the degree of regional connectivity are (a) the endowment, scarcity, and accessibility of resources within the connected region; (b) the financial capacity to activate these resources as production factors; and (c) the institutional conditions enabling resource transfers to occur.

Until 40 years ago the Mekong River remained a relatively unmodified river system of low impoundment, connecting the primary livelihood pursuits of agriculture, fishing and forestry of a predominately rural population. The Mekong has thus acted as an historical conduit of relatively stable cultural, economic, agricultural and spiritual connection across the Mekong Region countries, despite periods of political and economic turbulence (Molle et al. 2009).

A set of bio-physical and hydrological conditions over the past 40 years has also promoted high levels of economic and political connectivity between Mekong riparian countries (Molle et al. 2009; Theeravit 2003). The steep elevation gradients of the head waters and upper catchments of the main tributaries have provided

Table 1.1 Population (in million people) and area (in km^2) of the wider Mekong Region, 2011, sources from CIA factbook (https://www.cia.gov/library/publications/the-world-factbook/)

	Population million people	Area km^2
Thailand	66.7	513,000
Vietnam	90.5	331,210
Lao PDR	6.5	236,800
Cambodia	14.7	181,035
Myanmar	54.0	676,578
Yunnan	45.7	394,000
Mekong Region	278.1	2,332,623

opportunities for impoundments and hydropower generation, tentatively coexisting with biodiversity hotpots, small scale localised irrigation, and swidden agriculture. The rapid gradient transition to the extensive plains and deltas has allowed water diversions for agrarian landscapes, including extensive irrigation in the delta, fisheries, and river-based transport.

Human migration, natural resource flows, and financial investments are amongst a cluster of factors that influence the critical dynamics generating increased levels of connectivity between Mekong countries (Contreras 2007; Dore 2003; Harima et al. 2003; Theeravit 2003). Theeravit (2003) claims the financial strength of Chinese, Thai, and Vietnamese private and State companies has accelerated their potential to invest internationally. As a consequence, private and State investors are replacing traditional donor organisations in order to source natural resources or manufacturing capacity in neighbouring countries (Middleton et al. 2009; Molle et al. 2009). Resource endowments are however, geographically dispersed and characterised by variable extraction costs and relative scarcity at national scales. A political environment conducive to increased direct foreign and private investment is likely to re-align institutional arrangements with biophysical conditions, facilitating increased supply of demanded resources and stimulating wider Mekong Region connectivity.

The constellation of biophysical and socio-economic factors and dynamics has implications for the political economy of the Mekong (Table 1.1) (Dore 2003). Inter country discussions to activate economic potential and satisfy aspirations of national economic growth can stimulate the development of supra national institutional arrangements and governance processes that reinforce regional connectivity.

Examples for such reinforcing processes are bilateral agreements to reduce the capital outlays and operational costs of infrastructure necessary for transporting resources. Changes in land access rights and foreign investment regulations are likely institutional amendments. As a corollary, changes in migration legislation might be required to satisfy labor requirements, further buttressing the connectivity beyond national boundaries. Similar implications arise for the relationships between labor, electricity, water, minerals, and agricultural resources. As these links intensify, regional connectivity increases and over time, a highly connected Mekong Region can emerge.

The portfolio of instruments, responses, strategies, and metrics developed in the wider Mekong Region is generally associated with relatively stable and weak levels of connectivity associated with largely unmodified rivers, such as the Mekong, the Irrawaddy and the Salween. As is the expertise, experience, and evidence relied on to adjudicate and evaluate impeding investment decisions. Periodic amendments have arisen, but have not been subject to notions or threats of redundancy and high probabilities of policy failure due to changed connectivity levels. A tension arises when institutional arrangements and analysis underpinning policy decisions in connected regions are geared to assumptions of weak connectivity and analysis confined to a single economic sector (De Landa 2006; Molle et al. 2009).

Once a region is highly connected, some decisions are likely to have implications beyond their national or sub-national intention. Improving the understanding of these regional implications was the primary aim of the expert panel assessment documented in this book.

2 The Water-Food-Energy Nexus

Besides regional connectivity, this assessment considers also cross-sectoral implications, in particular between the water, food and energy sectors. The majority of planned and implemented development decisions in the wider Mekong Region aim for either improved water access, increased energy supply or improved food security. Investments in any of these three sectors seem critical as these sectors are closely linked, harbouring potential trade-offs and unintended side effects.

The Bonn2011 conference on the Water, Energy and Food Security Nexus, argued that improved understanding of the dynamics and linkages between these sectors is crucial to understanding potential opportunities, trade-offs and synergies, and thus to avoid mal-adaption and ineffective investments. This implies, they state, that "Understanding the nexus is needed to develop policies, strategies and investments to exploit synergies and mitigate trade-offs among these three development goals with active participation of and among government agencies, the private sector and civil society. In this way, unintended consequences can be avoided." (Bonn2011, 2011) This statement emphasises the importance of this Nexus for the development context. Figure 1.2 shows key processes in Nexus dynamics.

Acknowledging these potential trade-offs and synergies makes some of the risks explicit when addressing sectoral goals in isolation. Investments that pursue energy goals could create substantial trade-offs or synergies. Trade-offs might offset expected impacts of investments in the food security space or vice versa. Coordinating investments and developing consistent policies that allow for sustainable development would ideally involve:

(a) The understanding of these relevant connections;
(b) Specifying potential trade-offs and synergies for the particular context;
(c) The design of effective measures that help mitigating trade-offs and exploiting synergies; and
(d) Monitoring and assessing impacts of investments on Nexus dynamics.

It is crucial to understand the Nexus as a dynamic system; investments in this Nexus are likely to change or even transform the nature of some or all links, which emphasises the importance of monitoring and assessing impacts.

However, methodologies are limited. The development of analytical tools, methodologies and empirical processes are even more challenging for the water-food-energy Nexus (Smajgl et al. in review).

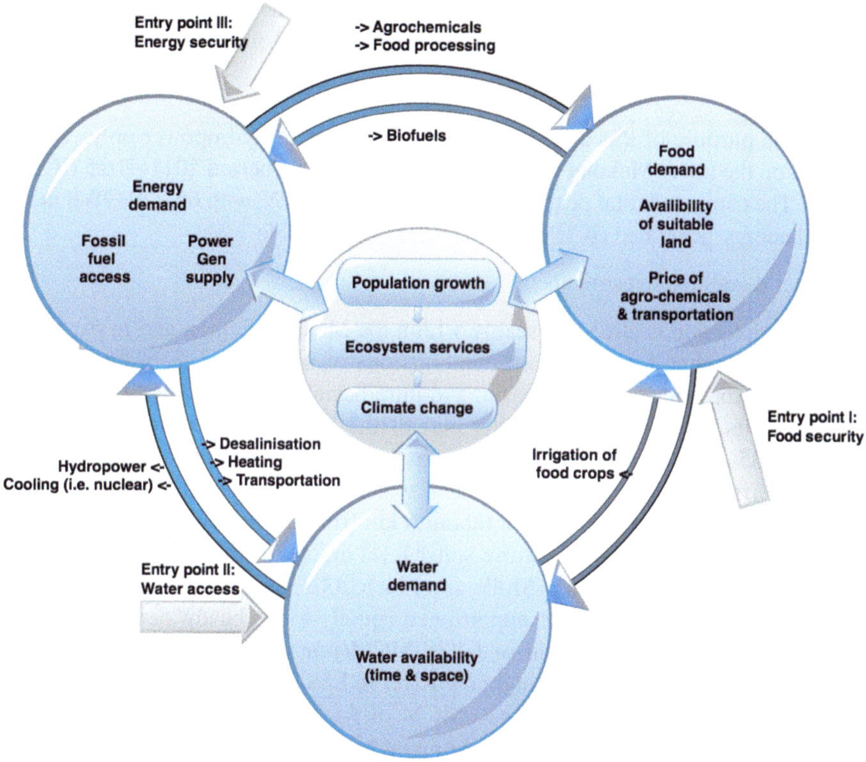

Fig. 1.2 The Water-Food-Energy Nexus

3 Decisions with Regional Implications

For this assessment, two stakeholder workshops identified a list of six large-scale national investments with the potential of having regional implications. They include

- Mainstream dams in the lower Mekong basin, in particular in Lao PDR and Cambodia
- Large scale water diversion, in particular in Thailand and Lao PDR
- Investments in response to sea level rise
- Land use changes in response to accelerated increases in rubber demand
- Construction of transport infrastructure, in particular the proposed Kunming-Phnom Penh railway
- Mining operations, in particular bauxite mining in southern Lao PDR, northeast Cambodia and southwest Vietnam

All six decisions are likely to have implications beyond national borders, highlighting the importance of assessing these investments. As these decisions are contemplated and not yet implemented. Therefore, we need to make assumptions for the expert panel assessment, which we lay out in the following.

- Lower Mekong River mainstream dams and projects

For the purpose of this assessment we assume that 12 hydropower projects will be built on the lower Mekong mainstream during the time period 2011–2025 (ICEM 2010). The estimated total peaking capacity is 12,980 MW, with 64,229 GWh mean annual energy generated (ICEM 2010) (ICEM 2010: Table 1.2).

- Water diversion investments

Water diversion is already a reality within the Mekong Region and in particular within the Mekong basin. Several combinations of inter-basin, from the Mekong River and tributaries out of the basin, and intra basin diversions, from the Mekong River and its tributaries into another part of the Mekong basin were considered (MRC 2005a, b). For the purpose of this assessment we assume intra basin diversions for irrigation in north east Thailand, which includes diversion of water from the Lao PDR tributary Nam Ngum (about 6 km from the confluence: see Fig. 1.3) by building a diversion dyke to raise water level high enough for gravity diversion into tunnel under the Mekong (Molle and Floch 2008). We assume a constant rate of 300 m^3/s for the whole year, apportioned equally to the headwaters of the Mun River and Chi River (or a total of 6,878 MCM/year). This assumption includes building two dams on the Xe Banghiang River in Lao PDR, close to the confluence with the Mekong, from which 3,320 MCM of water (or 150 m^3/s) could also be abstracted and siphoned under the Mekong into Isaan (MRC 2005a).

- Sea-level rise

Following Rahmstorf (2007) global sea levels are likely to increase assuming Intergovernmental Panel on Climate Change (IPCC) assumptions on increasing global temperature. Average projections estimate global sea level rise at around 200 mm by 2030 (Rahmstorf 2007). The Vietnamese Institute for Meteorology, Hydrology and Environment assumes 65–100 cm by 2,100 (IMHEN 2010, p. 30). Similarly, the IPCC predicts sealevel rise for the Mekong Region of up to 0.59 m by 2100, without consideration of reduced global ice stocks and subsequent ice melt (IPCC 2007, Sect. 5.2), also referred to by the Mekong River Commission (MRC 2009). Due to the shorter time horizon of our assessment we refer to Rahmstorf (2007) and assume a rise of 20 cm by 2030, which corresponds with Wassmann et al. (2004).

We are lacking the adequate resolution of a digital elevation data for the wider Mekong Region to calculate the exact area inundated by such a sea level rise. Although most available simulations do not allow for assumptions below 1 m (an exception can be found at http://globalfloodmap.org/Vietnam) it seems likely that an increase of 200 mm will not directly inundate substantial areas. Wassmann et al. (2004) assume an increase of 200 mm for their simulation, (see Fig. 1.4). However, storm surge is likely to increase and the rate of salinity intrusion and coastal erosion accelerated. Therefore, we assume that substantial parts of Vietnam's Mekong delta will experience more severe storm surge damage (Depicted in red in Fig. 1.4). Salinity intrusion is already a significant problem in the Mekong delta as depicted in Fig. 1.5 (MRC 2003). We assume that a sea level rise of 200 mm will further accelerate this problem.

Table 1.2 Assumptions for the mainstream dam construction (Source: ICEM 2010: Table 1.2. p. 34)

Mainstream dam			Management status			Design specifications									Dimensions		
	Location	Developer	Earliest potential commission date	Design status	Environmental assessment status	Rated head (m)	Plant design discharge (m^3/s)	Installed capacity (MW)	Peaking capability (MW)	Mean annual energy (GWh)	Firm annual energy (GWh)	Full supply level (mamsl)	Low supply level (Mamsl)	Live storage (mcm)	Reservoir area (km^2)	Length of dam (m)	Height (m)
Pak Beng	Lao PDR	Datang International Power Generation	2016	MoU, feasibility	IEE submitted	31	7,250	1,230	1,230	5,517	4,073	340	334	442	146	943	76
Louang Prabang	Lao PDR	PetroVietnam Power Corporation	2016	MoU, feasibility	Feasibility study	40	3,812	1,410	1,412	5,437	4,205	310	308	734	89	1,106	68
Xayaburi	Lao PDR	SEAN & Ch. Karnchang Public Co Ltd	2016	MoU, feasibility	Feasibility and full ESIA submitted	24	6,018	1,260	1,260	6,035	5,139	275	270	225	61	810	32
Pak Lay	Lao PDR	CEIEC and Sino-Hydro	2016	MoU, feasibility	IEE submitted	26	4,500	1,320	1,320	6,460	4,252	240	237	384	67	630	35
Sanakham	Lao PDR	Datang International Power Generation	2016	MoU, feasibility	Not yet	25	5,918	700	1,200	5,015	3,978	220	215	106	70	1,144	38
Pakchom	Lao PDR	N/a	2017	MasterPlan	Not yet	22	5,720	1,079	1,079	5,318	5,052	192	190	12	108	1,200	55
Ban Koum	Lao PDR	Italian Thai Asia Corp. Holdings	2017	MoU, feasibility	Not yet	19	11,700	1,8720	1,8720	8,434	8,012	115	115	0	234	780	53
Latsua	LaoPDR	Charoen Energy and Water Asia Co Ltd	2018	MoU, pre-feasibility	Pre-feasibility study submitted	10.6	10,000	686	686	2,668	1,524	97.5	95.5	0	10	1,300	27

(continued)

Table 1.2 (continued)

Mainstream dam			Management status			Design specifications									Dimensions		
	Location	Developer	Earliest potential commission date	Design status	Environmental assessment status	Rated head (m)	Plant design discharge (m^3/s)	Installed capacity (MW)	Peaking capability (MW)	Mean annual energy (GWh)	Firm annual energy (GWh)	Full supply level (mamsl)	Low supply level (Mamsl)	Live storage (mcm)	Reservoir area (km^2)	Length of dam (m)	Height (m)
Don Sahong	Lao PDR	Mega First	2016	PDA, detailed planning	Full EIA submitted, additional studies	17	2,400	240	240	2,375	1,989	75	72	115	3	1820-720 -2730	10.6-8.2-8.3
Thakho diversion	Lao PDR	CNR & EDL	2016	MoU, pre-feasibility	IEE submitted	16	380	50	50	360		71.7	68.7	N/a	N/a	Channel – 1,800 m	N/a
Stung Treng	Cambodia	Vietnam Urban and Industrial Zone Development Inv. Corp	N/a	MoU, pre-feasibility	Not yet	15	18,493	980	591	4,870	2,937	55	50	70	234	10,884	22
Sambor	Cambodia	China Southern Power Grid	2020	MoU, pre-feasibility	Pre-feasibility submitted	33	17,668	2,600	2,030	11,740	9,150	40	39	465	716	18,002	56

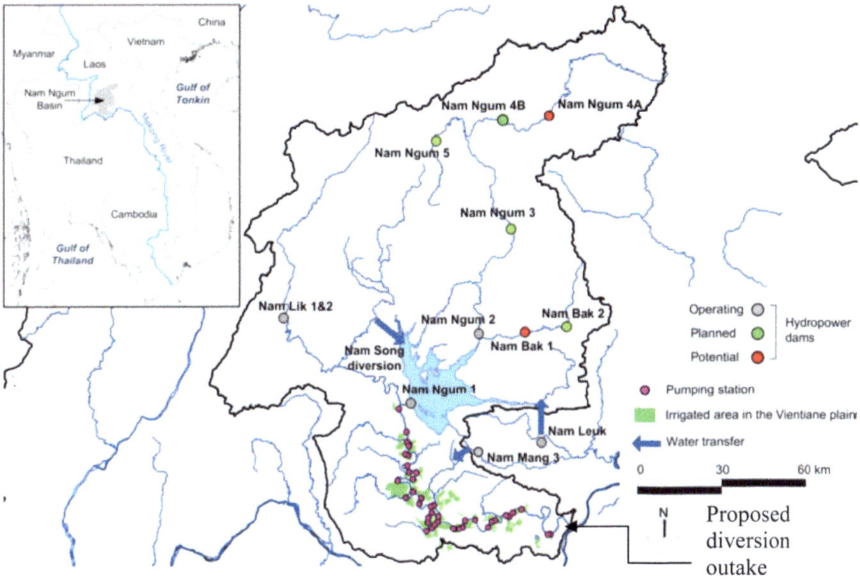

Fig. 1.3 Location of hydropower dams, pumping stations and proposed water diversion in the Nam Ngum basin (Source Lacombe et al. (2012))

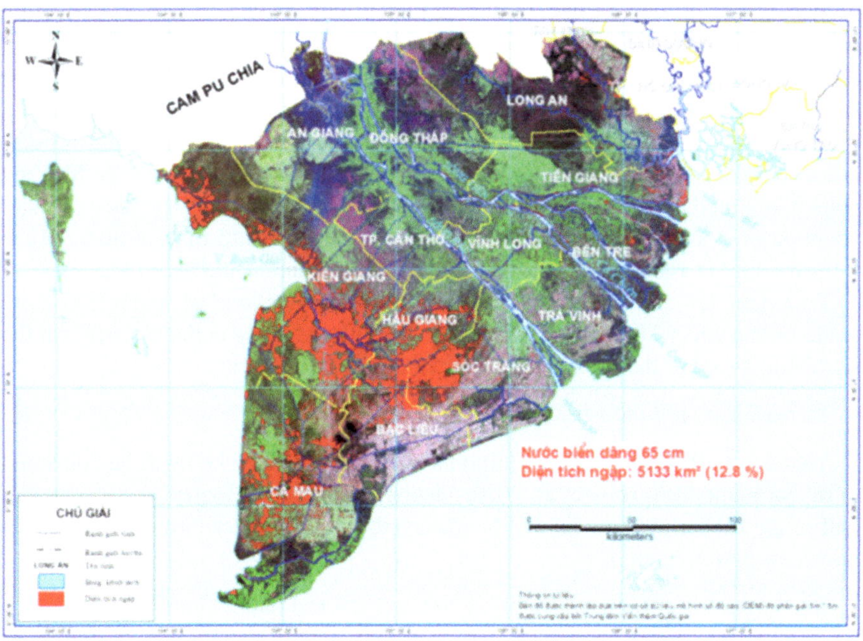

Fig. 1.4 Sealevel rise impacts on the Mekong delta depicting in red increasing inundation (Source: Wassmann et al. (2004))

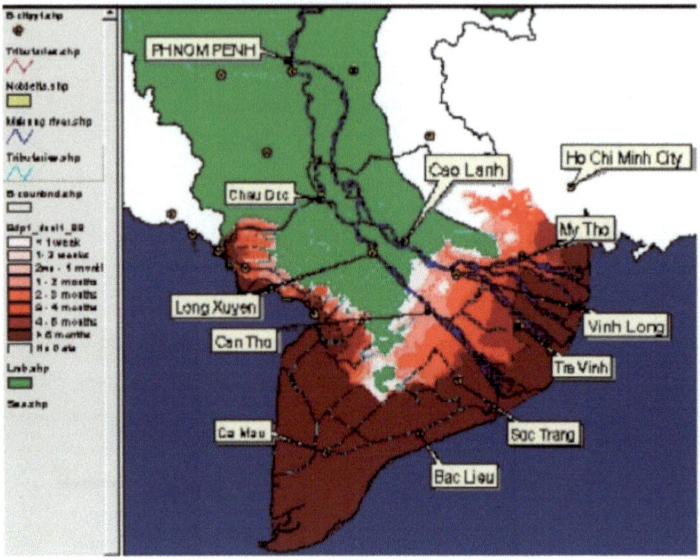

Fig. 1.5 Mekong delta – simulation of saline intrusion during the dry season drought conditions of 1998. The map shows the duration of salinity levels greater than 1 g/l. The area affected exceeds half of the total 55,000 km^2 that defines the main delta (Source MRC 2003, State of the basin report)

We assume that the Vietnamese government will invest in water infrastructure (dikes, sluice gates) to offset any direct inundation. We also assume that accelerated salinisation and increasing damage by storm surges are unlikely to be avoided. Most relevant adaptation mechanisms are assumed to occur at the household level by changing to alternate crops or other land uses. Exploring alternate farming systems is supported by the central government as the rice production goal for the Mekong delta is assumed to be revised before 2030. Incentives are currently in place to foster adoption of salt resistant rice varieties and integrated shrimp-rice farming. Urban infrastructure is assumed to be protected and mainly unaffected by 2030.

Important for this assessment is that similar implications of sealevel rise are made for the wider Mekong Region, not only for the Mekong delta. In particular the Cambodian side of the Mekong delta and the Red River delta.

- Increased area of industrial rubber plantations

We assume that by 2050 an additional area of 1.6 millions ha will be converted to rubber plantations (Fox et al. 2012; Ziegler et al. 2009). Further we assume that half of the rubber plantations will be managed by smallholder farmers and half by large concession holders.

- Railway connection from Kunming to Lao PDR to Phnom Penh

Regional railway lines in all countries will be developed (or rehabilitated) to further regional integration and connectivity in the Mekong Region (Fig. 1.6)

- Cambodia: rehabilitation of the railway funded by Asian Development Bank (ADB) and AusAID

Fig. 1.6 Railway lines under construction in the Mekong Region

- China is financing the Feasibility Study for line between Phnom Penh and VN border
- China is building new line to the VN and Myanmar borders
- Thailand considering may develop a high-speed train line and new links to link Lao and onward to VN
- VN considering lines to Lao and Cambodia

Plans for realising such railway links are accelerating due to escalating sub-regional trade, growing concerns over climate change and more recently – fluctuating fuel costs.

This assessment assumes that the railway through Lao PDR will be in construction by 2015 and form part of the Asian-China railway, which might run one day from Yunnan Province south through Lao PDR to Thailand, Malaysia and Singapore.

- Bauxite mining

Prospecting for commercial deposits of bauxite has been conducted for the last 5 years in the Bolaven plateau in southern Lao PDR (Champassak Province next to the Cambodian border). The volume of extracted bauxite necessary to ensure commercial viability translates into large mining areas, the majority of which will have to be cleared of vegetation. In a review of seven Bauxite concessions located in the Bolaven Plateau, Lazarus (2009) estimates the aggregate area conceded in 2009 at 1,250 km^2, with expansions of a further 950 km^2 in negotiation. Total bauxite deposits are estimated at more than three billion tonnes (Fig. 1.7).

To make alumina refinery commercially viable at least 0.5 million tons of alumina output are required or between 1.0 and 1.5 million tons of raw bauxite per year, dependent on the grade of the bauxite, for at least 20 years. Such a venture needs a considerable volume of electricity, possibly in the region of 150 MW and large quantities of water (Sekong River and Sekong dams).

We assume that the processing of alumina into a smelting operation producing aluminium is not an option, as it requires 600–800 MW of 100 % reliable, base load electricity. Electricity demand is dependent on the technology applied and the scale of the operation. It would only be possible if more hydropower development becomes available or construction of suitably sized fossil fuel fired generator.

4 Methodology

Imperative to this study are two generic assumptions. First, we assume that the implementation of multiple development investments will have diverse and multiple consequences greater than the sum of individual initiatives. This assumption is critical to the concept of systems emergence and a theoretical understanding of complex systems. Outcomes emerge from the combination of different interventions and cannot be reduced to individual effects. Irreducibility dictates that outcomes and individual effects are also non-commensurate. Understanding the emergent properties of the combination of various contemplated investment strategies requires a systematic cumulative assessment.

Second, transboundary implications within a highly connected region are likely to impact on the connecting drivers, such as human migration, natural resources or financial flows. For instance, sectoral investments in one location can impact on downstream water availability or quality. Thus, downstream conditions for households might deteriorate to a point where households may be forced or prefer to migrate. If migration coincides with improved conditions in the locality of the initial investment affected households might be attracted to move into this upstream area. Such feedback loops can substantially and rapidly alter the initial conditions confronting decision makers and their ability to achieve investment goals. An amended set of initial conditions implies that the potential for unexpected side effects increases with increased connectivity. Such increasing complexity needs to be considered to improve the understanding of likely emergent effects of large scale decisions such as the six impending investments considered in this Volume.

Both assumptions pose a methodological challenge, which we address by deploying a mixed method approach, including participatory scenario building, agent-based social simulation, and a highly structured expert panel approach (Smajgl et al. in review; Smajgl and Ward in review). This Volume is focused on presenting the results from the Delphi technique-expert panel approach.

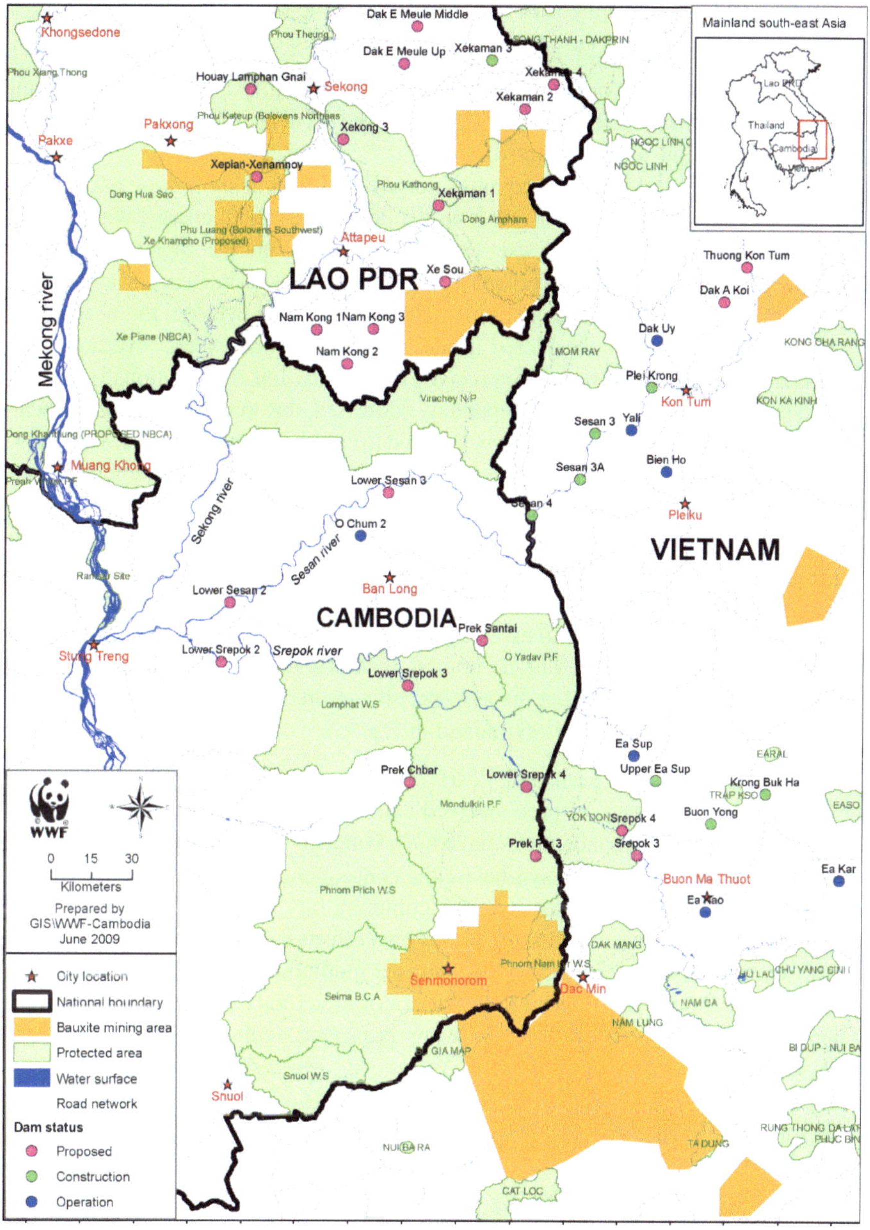

Fig. 1.7 Bauxite deposits and mining claims in Lao PDR, Cambodia and Vietnam (Source: WWF (2008))

Expert panels or Delphi techniques are commonly defined as "a method for structuring a group communication process so that the process is effective in allowing a group of individuals, as a whole, to deal with a complex problem" (Linstone and Turoff 1975, p. 3). Many variations of the principle Delphi technique have been applied and tested. In this case the following steps were conducted:

First, a stakeholder workshop[1] was conducted to identify the national decisions that are likely to have regional implications. The six selected impending development investments are listed above.

Second, the workshop participants selected those sectors most critical for understanding regional implications, ranking the domains of food, energy, water, livelihoods, migration, and land use change as those with the highest relevance. Participants agreed that a prerequisite of contracted experts was an ability and expertise to contribute regional (not only national) knowledge for a specified domain. However, we considered it beneficial to find one expert covering livelihoods and migration as both are perceived as critically linked. Additionally, mining and land use change were to be separately addressed, due to the magnitude of localised mining effects. Six experts were selected to cover the selected sectoral domains.

As a third step, each expert was asked to conduct a desktop assessment of each of the six individual investments followed by a cumulative assessment assuming all six changes would simultaneously occur. The two-staged perspective was employed to conduct all six sectoral assessments. Chapters 2, 3, 4, 5, 6, and 7 detail the assessments, synthesising available literature, data and expert opinion.

Finally, the cross-sectoral assessment commenced with a structured and facilitated workshop comprised of six invited experts and ten additional regional and agency experts ensuring that the assessment captured a broader perspective. The structure of the workshop is explained in Fig. 1.8.

Several scholars (Halpern and Pearl 2005; White 1998; Zaleski 1988) have pointed out the relevance of preconceived causal relationships, in particular those among researchers. Preconceptions have the advantage that experienced experts are able to rapidly and effectively synthesise the consequences of key potential changes and are familiar with the most relevant contemporary evidence. However, when considering multiple investments, a few marginal effects can accumulate and manifest as substantial impacts. Consequently, a revised methodology challenged experts to think outside deeply 'entrained' cause-effect storylines and confront existing biases. Therefore, as a first phase, experts were presented with the suite of sectoral assessments.

Second, they were only confronted with the effects of development investments, independent of proposed causal factors and determinants.

Third, the effects systematically listed within sectoral assessments were presented and experts asked to identify likely sectoral reactions and events that would be activated by these changes. Interestingly, discussions repeatedly returned to the question of what caused individual effects, highlighting a predisposition to return to established explanations. However, this process allowed for an effective interaction and intersecting of sectoral assessments. As Fig. 1.8 shows the process was repeated

[1]Workshop participants (10–14 January 2011): Lilao Bouapao, Thierry Facon, Tira Foran, Kate Lazarus, Robert Mather, Andrew Noble, Lu Xing, Juha Sarkkula, Pech Sokhem, and Pham Quang Tu.

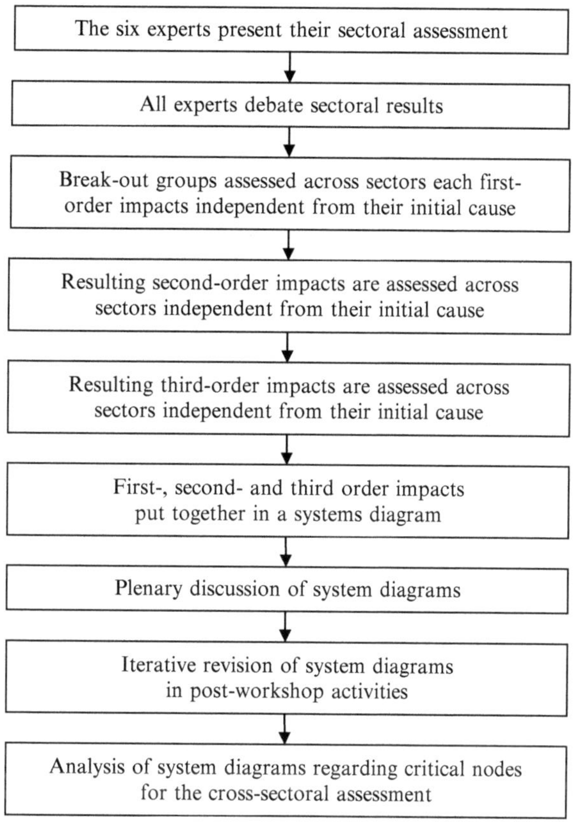

Fig. 1.8 Expert panel process

to specify second and third order impacts. Finally, cause effect-chains were constructed that were independent of individual sectoral perspectives and arranged in system diagrams. During the final step of the workshop and more importantly in a post-workshop process, these system diagrams were refined based on recorded workshop deliberations (see Annex A-F).

The final step for conducting the cross-sectoral assessment included the analysis of all six system diagrams, one for each chosen investment. Diagrams were analysed regarding critical storylines and key variables. Additionally, the six diagrams were overlayed to identify critical processes for cumulative, cross-sectoral outcomes of these six changes. Results from both levels of analysis of this cross-sectoral assessment are summarised in Chap. 8.

5 Methodological Limitations

The Delphi technique and specifically expert panel assessments necessarily rely on individual opinions and evaluations. Linstone and Turoff (2002) p. 6, note however that it is an analytical technique that "*can benefit from subjective judgements on a collective basis*." The expert panel methodology deployed in the project (illustrated in Fig. 1.8) introduced structured, facilitated deliberation; iterative feedback amendments based on meeting discussions and; information refinements of the initial expert appraisals. These features are consistent with the main process recommendations noted by Linstone and Turoff (2002), intended to assist the panel to identify and address subjective biases and promote more impartial assessments.

Therefore, the resulting sectoral assessments presented in this volume retain a residue of bias and document the opinions of the individual expert. The respective Chapters include the selective inclusion and omission of literature based insights and evidence experts considered relevant or crucial to their specific sectoral assessment. As editors we considered that additional editorial interventions by way of analytical and literature guidance would have introduced a level of censorship sufficient to compromise the Delphi protocols. It is noteworthy that the resulting chapters also reflect prevailing biases and preconceptions that dominate the current suite of debates and advisory processes concerned with development investments in the wider Mekong region.

In addition to existing expert influences and predispositions, the communication of cross-sectoral impacts of multiple development investments poses a formidable challenge, compounded by the transboundary context of the wider Mekong Region. The following Chapters reflect a specified sectoral focus guided by the project imperative to evaluate the singular sectoral ramifications (and iterative degrees of connectivity) of the six prescribed development investment decisions. As a corollary, the reader of this Volume may be left with the impression of an unbalanced assessment when reading individual Chapters. This is a necessary consequence of applying singularly focussed sectoral perspectives, reflecting the topics and diverse values contested within particular interest groups. A development strategy might look entirely negative or entirely positive when assessed from the perspective of one particular sector. For instance, a development strategy negatively assessed when viewed through the lens of food security may conversely confer substantial benefits in the energy domain.

For example, the Chapter by Sokhem Pech emphasises the effect on the water sector of a modified Mekong hydrology from proposed dam impoundments, coupled with a more cursory appraisal of the benefits of energy generation accruing to Laos and Cambodia. Whilst increased dry season flows are one dimension of dam operations, reduced wet season flows and the attendant decreases in the mean amplitude of wet/dry season fluctuations also warrant consideration. Alternatively, Tira Foran's Chapter on the Mekong energy system details a focussed evaluation of the potential effect on the energy sector of proposed Mekong water developments, including dams and increased irrigation abstractions for the production of biofuels.

Thus, individual Chapters can be misinterpreted as biased as they convey a particularly negative or positive appraisal. In a general sense, the Chapters forecast negative consequences of proposed development investments, accentuated by the cumulative evaluations. However, we encourage the Reader to consider that a critical interpretation of expressed opinions across the Chapters is likely to promote a more balanced view. This reflects the rationale of this Volume, as we aim to portray the manifold sectoral outcomes of the impending development investments. Based on the evidence and opinions expressed in the Volume, we argue that such diverse outcomes warrants the implementation of comprehensive cross-sectoral assessments that account for geographical trade-offs in the Mekong transboundary context. Sen (1995, 2009), argues that identifying anticipated intervention trade-offs, coupled with transparent and rigorous public reasoning, negotiation and external mediation is the most effective paradigm for achieving equitable and sustainable development.

Another methodological limitation was the availability of reliable, formally compiled data for the wider Mekong Region system. While we argue that the high connectivity of the wider Mekong region demands an additional cross scale assessment of development strategies in the regional context to reveal unintended side effects, accessible data is not available for all locales. Consequently, the following Chapters are geographically biased towards the Lower Mekong Basin as it is there where most studies have been carried out. Additionally, some of the references cited by Chapter authors are characterised by slightly different geographical definitions, introducing the potential for confusing spatial foci. However, the intent was an extensive consideration of as much available evidence for the wider Mekong Region, including partial analysis. Alternatively, the literature based reviews and assessments would be severely constrained. We hope that the inconsistency and partiality in geographic foci within the wider Mekong region does not distract the reader from gaining an improved understanding on how development decisions might play out for different sectors and different countries in the wider Mekong region.

References

Bonn2011. 2011. Messages from the Bonn2011 conference: The water, energy and food security nexus – Solutions for a green economy.

Cechich, A., M. Piattini, A. Vallecillo, C. Atkinson, C. Bunse, and J. Wust. 2003. Driving component-based software development through quality modelling. In *Component-based software quality*, 207–224. Berlin/Heidelberg: Springer.

Contreras, A. 2007. Synthesis: discourse, power and knowledge. In *Democratizing water governance in the Mekong Region*, ed. L. Lebel, J. Dore, R. Daniel, and Y.S. Koma, 227–236. Chiang Mai: Mekong Press.

De Landa, M. 2006. *A new philosophy of society: Assemblage theory and social complexity*. New York: Continuum International Publishing Group.

Dore, J. 2003. The governance of increasing Mekong regionalism. In *Social challenges for the Mekong Region*, ed. M. Kaosa-ard and J. Dore, 405–440. Bangkok: White Lotus.

Fox, J. M., J. B. Vogler, O. L. Sen, A. L. Ziegler, and T. W. Giambelluca. (2012), 'Simulating Land-Cover Change in Montane Mainland Southeast Asia', *Environmental Management*, 49 (5), 968–79.

Halpern, J. and Pearl, J. (2005), 'Causes and Explanations: A Structural-Model Approach, Part I: Causes.', *British Journal of Philosophy of Science*, 56 (4), 843–87.

Harima, R., R. Varona, and C. DeFalco. 2003. Migration. In *Social challenges for the Mekong Region*, ed. M. Kaosa-ard and J. Dore, 225–261. Bangkok: White Lotus.

Harold, A. L., and Murray. Turoff, eds. 2002. *The Delphi method: Techniques and applications*. Electronic version http://is.njit.edu/pubs/delphibook/. Accessed 24 Oct 2012.

ICEM. 2010. MRC strategic environmental assessment of hydropower on the Mekong mainstream. Impacts assessment (opportunities and risks) discussion draft. 14 May 2010. Hanoi: International Centre for Environmental Management.

IMHEN. 2010. *Impacts of climate change on water resources and adaptation measures*. Hanoi: Vietnam Institute of Meteorology, Hydrology and Environment.

IPCC. 2007. *Fourth assessment report: Climate change 2007*. Cambridge, UK/New York: Intergovernmental Panel on Climate Change.

Lacombe, G., Smakhtin, V., and Hoanh, C. T. (2012), 'Wetting tendency in the Central Mekong Basin consistent with climate change-induced atmospheric disturbances already observed in East Asia', *Theoretical and Applied Climatology*, 1–13.

Lazarus, K. 2009. *In search of aluminum: China's role in the Mekong Region*. Cambodia: Heinrich Böll Stiftung Cambodia, World Wild Fund and International Institute for Sustainable Development.

Linstone, H.A., and M. Turoff. 1975. Introduction. In *The Delphi method: Techniques and applications*, ed. H.A. Linstone and M. Turoff. Reading: Addison-Wesley.

Middleton, C., J. Garcia, and T. Foran. 2009. Old and New hydropower players in the Mekong Region: Agendas and strategies. In *Contested waterscapes in the Mekong Region*, ed. F. Molle, T. Foran, and M. Käkönen, 23–54. London: Earthscan.

Molle, F., and P. Floch. 2008. *The "desert bloom" syndrome: Irrigation development, politics, and ideology in the northeast of Thailand*. Chiang Mai: Institut de Recherche pour le Développement, International Water Management Institute.

Molle, F., T. Foran, and P. Floch. 2009. Changing waterscapes in the Mekong Region: Historical background and context. In *Contested waterscapes in the Mekong Region: Hydropower, livelihoods and governance*, ed. F. Molle, T. Foran, and M. Käkönen, 1–21. London: Earthscan.

Mekong River Commission (MRC) (2003). State of the Basin Report. MRC: Phnom Penh.

MRC. 2005a. *The MRC basin development plan – scenario for strategic planning, BDP library*, vol. 4. Vientiane: Mekong River Commission Secretariat.

MRC. 2005b. *Overview of the hydrology of the Mekong basin*. Vientiane: Mekong River Commission Secretariat.

MRC. 2009. *Adaptation to climate change in the countries of the lower Mekong basin*. Vientiane: Mekong River Commission (MRC).

Rahmstorf, S. 2007. A semi-empirical approach to projecting future sea-level rise. *Science* 315: 368–370.

Sawyer, R.K. 2005. *Social emergence: Societies as complex systems*. New York: Cambridge University Press.

Sen, A. 1995. Rationality and social choice 900. *American Economic Review* 85: 1–24.

Sen, A. 2009. *The idea of justice*. Cambridge, MA: The Belknap Press/Harvard University.

Smajgl, A., T. Foran, J. Dore, J. Ward, and S. Larson. In review. Visions, beliefs and transformation: Methods for understanding cross-scale and trans-boundary dynamics in the wider Mekong Region. *Global Environmental Change*.

Smajgl, A. and Ward, J. (in review), 'A design protocol for research impact evaluation: Development investments of the Mekong region', *Research Evaluation*.

Theeravit, K. 2003. Relationships within and between the Mekong Region in the context of globalisation. In *Social challenges for the Mekong Region*, ed. M. Kaosa-ard and J. Dore, 49–80. Bangkok: White Lotus.

Wassmann, R., N.X. Hien, C.T. Hoanh, and T.P. Tuong. 2004. Sea level rise affecting the Vietnamese Mekong delta: Water elevation in the flood season and implications for the rice production. *Climate Change* 66: 89–107.

White, Peter A. (1998), 'Causal Judgement: Use of Different Types of Contingency Information as Confirmatory and Disconfirmatory', *European Journal of Cognitive Psychology*, 10 (2), 131–70.

WWF. 2008. *Livelihood sustainability analysis in Mondulkiri province*. Phnom Penh: WWF Greater Mekong-Cambodia Country Program.

Zaleski, Z. (1988), 'Attributions and emotions related to future goal attainment', *Journal of Educational Psychology*, 80 (4), 563–68.

Ziegler, A.D., J.M. Fox, and J. Xu. 2009. The rubber juggernaut. *Science* 324: 1024–1025.

Chapter 2
Water Sector Analysis

Sokhem Pech

1 Introduction

This Chapter focuses on the water resources and associated environs of the Mekong River Basin. First, the Chapter reviews and analyses the suite of potential implications for water arising from the implementation of the six contemplated decisions described in Chap. 1. Second, a derived cumulative analysis is described.

The Asian Development Bank (ADB 2009) divides the water sector into:

(1) Rural water – referring to aspects of rural water supply and sanitation, irrigation and drainage;
(2) Urban water – referring to urban water demands and water supply, sanitation and wastewater services, and urban environmental improvement; and
(3) Basin water – referring to the state of river health, planning, infrastructure including hydropower impoundments, natural hazard management, climate change, water catchment and wetland conservation.

The Mekong River Commission (MRC) classifies water as:

(1) Active sectors – referring to water supplies (domestic and industrial), irrigated agriculture, hydropower, and flood management and mitigation; and
(2) Passive sectors – referring fisheries, navigation and river works, tourism and water related recreation, and riverine environments (MRC 2010b).

Both the ADB and MRC water sector typologies have advocates and critiques, with for example the water sector differentiated into active and passive sectors strongly criticized by staff members of the MRC Secretariat (Regional POE 2010). The Chapter review adapted the water sector definition by the MRC, by considering an expanded

A. Smajgl and J. Ward, *The Water-Food-Energy Nexus in the Mekong Region: Assessing Development Strategies Considering Cross-Sectoral and Transboundary Impacts*,
DOI 10.1007/978-1-4614-6120-3_2, © Springer Science+Business Media New York 2013

set of relevant water use activities (including in-stream uses and other values) as equally important and active.

Where data of sufficient reliability is available, the Chapter discusses investment decision implications for the wider Mekong Region. However, the majority of available data and analyses for the water sector within the Mekong Region are concentrated on the Mekong basin.

The present Chapter contains three main sections:

Section 1. The Status Quo describes the current status of the 'water resources' sector in the wider Mekong Region and is intended to convey to the non-expert reader sufficient insight to appreciate the current sectoral perspective.

Section 2. An assessment of the six development decisions assuming they were to be implemented independently, highlighting the implications of each of these changes/decisions for Mekong basin water resources.

Section 3. A cumulative assessment of water resource outcomes, assuming the joint and concurrent implementation of the six investment decisions.

2 Status Quo of Water Resources in the Mekong Region

The Mekong Region's rich but fragile natural resources include a substantial and diverse agricultural base, timber and fisheries resources, considerable mineral potential, and extensive energy resources in the form of hydropower and large coal and petroleum reserves (ICEM 2010; ADB and SEI 2002). A number of development initiatives that effect water to varying degrees are at various stages of planning and development in the Mekong Region including; transportation (road transport, rail transport, water transport and air transport), energy generation (hydropower, natural gas), tourism, and international trade (ADB and SEI 2002; Pech and Sunada 2006; Molle et al. 2009).

The substantial and (until recently) relatively unmodified water resources of the Mekong basin have the capacity to support ongoing economic development in terms of irrigation and agricultural production, fishery and aquaculture, energy and forest products, navigation and other modes of transport, domestic and industrial water supply, and tourism (MRC 2010a).

2.1 Water Resources in the Mekong Region

Compared to many other global regions, the Mekong Region has high annual rainfall, especially in the mountainous catchments (see Table 2.2). Low annual precipitation amounts of less than 1,000 mm per year are found only in parts of Thailand, and at high elevation in the river headwater areas in Tibet, People's Republic of China (ADB and SEI 2002; MRC 2010a).

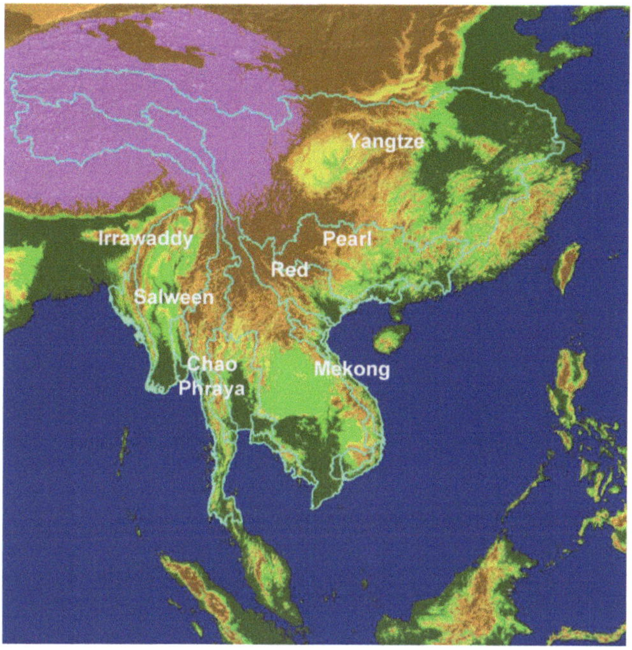

Fig. 2.1 Map of Mekong Region and major river systems (Source: Map courtesy of ICRAF)

The five major rivers in the Mekong Region depicted in Fig. 2.1 include: The Mekong (referred to as the Lancang Jiang in China), Red or Hong (Yuan Jiang), Chao Phraya, Irrawaddy and the Salween (Nu). Except for the Chao Phraya, the rivers are shared by more than two countries (ADB and SEI 2002). The mean volume of annual Mekong River runoff (475,000 million meter cubes (mm^3) in an average year) makes it the largest of the five Mekong Region rivers (Table 2.1) and the eighth largest river in the world (MRC 2010a). To date the Mekong River has also dominated development considerations and the subject of a diverse and competing set of claims. Snow melt in the upper Mekong catchment are the primary source of inflows from May to July, and summer monsoon rain from July to October are the main source of inflows (and floods) in both the main stem and tributaries, especially within Lao PDR and the Central Highlands of Viet Nam.

It is important to highlight that the upstream flows from the Upper Mekong in China contribute over 16 % of the total flow in an average year, while 55 % comes from the left bank tributaries in Lao PDR along with the Se Kong, Se San and Sre Pok (3S) River system (Vietnam Central Highlands, Lao PDR and Cambodia). However, during the dry season, snowmelt from China contributes 24.1 % of the total flows (see Table 2.2) (MRC 2010a).

Table 2.1 Summary of major river systems and reservoirs in the Mekong Region

Characteristic	Mekong	Red or Hong (Yuan Jiang)	Chao Phraya	Irrawaddy	Salween (Nu)
Countries in basin	All	PRC, Vietnam Lao PDR	Thailand	Myanmar, PRC, India	PRC, Thailand, Myanmar
Basin (catchment) area (km^2)	777,000 + 73,000 in Tibet & Qinghai	226,000 (40 % in Vietnam)	160,000	411,000	325,000
Area above 3,000 m altitude, km^2	62,000 (7 %)	Negligible	Zero	8,000 (<2 %)	90,000 (28 %)
Mean annual Runoff, mm	560	782	188	1,000	466
Water yield per annum, mm^3	475,000	177,000	29,800	410,000	151,000
Major dependent cities	Phnom Penh, Vientiane, Ho Chi Minh City, Can Tho, Khon Kaen	Hanoi, Hai Phong	Bangkok	Rangoon	Moulmein
Extent of lands under irrigation	Large	Large	Large	Large	Small
Number of existing storage reservoirs with surface area >20 km^2	xx	4	2	Data not avail	0
Level of water quality pollutants from human activities	Low (high at certain loclities during dry season low flow)	Medium	High	Low	Very low

Source: ADB and SEI (2002)

Compared to other global regions in the world in term of actual renewable water resources per capita,[1] the Mekong basin is not water stressed. The annual renewable water resources (ARWR) per capita give the maximum theoretical amount of water available per person, though in reality a large portion of this water may not be

[1] According to World Resources Institute, *Per Capita Actual Renewable Water Resources* is the maximum theoretical amount of water actually available on a per person basis for each country. It is a sum of internal renewable resources (IRWR) and external renewable resources (ERWR), including the flow for upstream and downstream countries and the potential reduction of external flow due to upstream water abstraction. Internal renewable water resources (IRWR) are comprised of the average annual flow of rivers and recharge of groundwater (aquifers) generated from endogenous (internal) precipitation. Even though IRWR measures a combination of surface and groundwater resources, it is typically less than the sum of the two because of overlap – water resources that are common to both surface and groundwater.

Table 2.2 Key hydrological characteristics of Mekong River Basin

	Yunnan	Myanmar	Lao PDR	Thailand	Cambodia	Vietnam	Total
Catchments (km2)	165,000	24,000	202,000	184,000	155,000	65,000	795,000
% of MRB total	22	3	25	23	19	8	100
% of total country's area	38	4	97	36	86	20	
Average rainfall (mm/year)	1,561		2,400	1,400	1,600	1,500	1,750
Average runoff (m^3/s)	2,414	300	5,270	2,560	2,860	1,660	15,060
Average runoff (MCM/year)	76,128	9,461	166,195	80,732	90,193	52,350	474,932
In dry season	19,032	1,419	24,929	12,110	13,529	7,852	78,871
Average runoff as % of total	16	2	35	17	19	11	100
In dry season	24.1	1.8	31.6	15.4	17.2	9.9	
Population (million)	10	0.5	4.9	24.6	10.8	21	71.8

Source: MRC (2005b), UNEP/GIWA (2006)

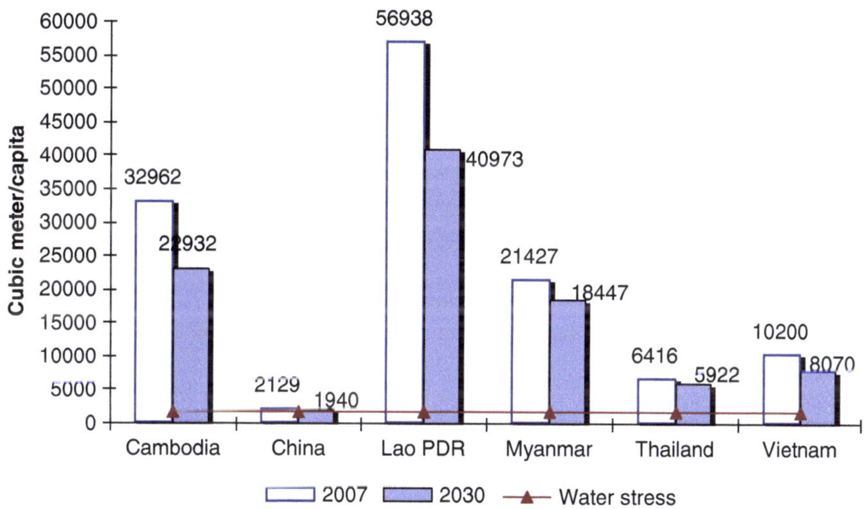

Fig. 2.2 Maximum theoretical amount of water actually available, on a per person basis, for each country in Mekong Region (provincial data for Yunnan not available) (Data source: WRI 2011)

accessible for human use (Ravenge and Mock 2000). National water stress is defined as annual per capita water availability below 1,700 m^3/year. In absolute terms, per capita water availability in Yunnan, Thailand and perhaps the Vietnam Delta are comparatively the lowest, whilst Lao PDR, Myanmar and Cambodia are well above the water stress limits (Fig. 2.2). Assuming that current water consumption patterns continue unabated, 20 year projections indicate that the most populous countries of

China, Vietnam and Thailand will tend towards increased water stress as water consumption rises (WRI 2011; see Fig. 2.1). Countries where the mean annual per capita water availability appears sufficient may actually face water shortages in the dry season and drought years. The highest marginal reduction in water availability per capita is estimated to occur in Cambodia and Lao (about 30 % of reduction), due to relatively high rates of population growth in the Mekong Region. It is important to note that the 2030 projections are slightly conservative because they are based on the UN's medium fertility assumption of population growth (UN 2007).

Competing water claims in the Mekong basin are closely related to the unequal spatial and temporal distribution of river flows, and the lack of a robust institution capable of negotiating and enforcing trans-national coordination and well-informed decision making for water resources development. Much of the runoff occurs during floods or is considered inaccessible because of remote locations. A proportional share of runoff is required to maintain other in-stream uses and non-consumptive social and ecological services (Halcrow Group 2003). For instance, the mean annual discharge of the Mekong River is 13,700 m^3/s with a peak wet season average discharge of 52,400 m^3/s resulting in widespread flooding. Minimum discharge during the dry season is approximately 30 times less at 1,600 m^3/s corresponding with the period of maximum water demand for food production (ADB and SEI 2002; MRC 2003).

Despite the relatively high renewable water resources per capita typifying Mekong basin countries, a number of locations currently face a series of critical water issues, such as:

- Water shortages in Thailand coupled with increasing irrigation water demands;
- Increasing salinity intrusion in the Vietnam Mekong delta;
- Threats and declines in basin fisheries and the degradation of natural habitats in many parts of the Basin;
- Recurring un-seasonal floods and drought;
- Reduced water quality, land-subsidence and morphological changes in delta areas; and
- Intensification of sectoral competition within and among the Mekong countries (MRC 2010a).

Based on the current hydrological reality of the Mekong River Basin, Fig. 2.3 describes a monthly assessment of water demands at key locations of River reaches.

2.2 *Water and Ecosystem Productivity and Integrity in the Mekong Basin*

The structure and functions of Mekong basin wetlands are closely linked to the seasonal flow pattern of the river, typified by a wet season flow of up to 10 m higher than the dry season. Fluctuations in river flow and consequent flooding change the

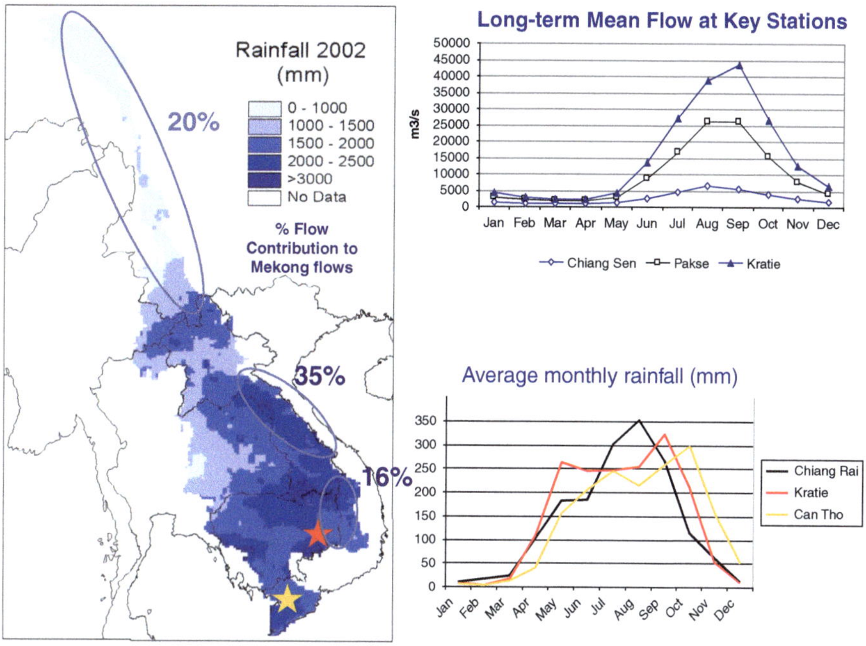

Fig. 2.3 Precipitation and flow contribution from MRB sub-basins (Source: MRC 2003)

structure and functionality of wetlands and subsequent productivity (Nikula 2008).[2] The river channel and wetland habitats are crucial for the ecological functioning of the river system.

The Mekong is a fluvial river system which changes substantially from upstream gradients to downstream deltas. Differences are evident as changes in water discharge and sediment transport (sediment flows, that are moved either on the bottom (bed load), in suspension, or in solution, or nutrients attached to the suspended matters, and/or in solution of the flows), and changes in the nature of the river geomorphology bed (bedrock, substratum, or geological base) (see e.g. Miyazawa et al. 2008). The fluvial continuum and hydro-system incorporates longitudinal exchanges of water, sediment, nutrients and species from upstream sources to the delta; lateral exchanges between the channel and its floodplain; and, vertical exchanges of water, nutrients and fauna between the river itself and the groundwater. These systems are of prime importance along the large rivers in the Mekong Region (Bravard and Goichot 2010).

The life-cycles of many Mekong fish species and adjacent coastal zones correspond to the cyclical annual fluctuations of the river's hydrological and sediment/

[2] An important feature is the Mekong Basin is its rich riverine ecology, fueled by the annual "flood pulse" especially in the Tonle Sap Great Lake where the seasonal cycle of changing water levels at Phnom Penh results in a very large flow reversal of water into and out of the lake via the Tonle Sap River.

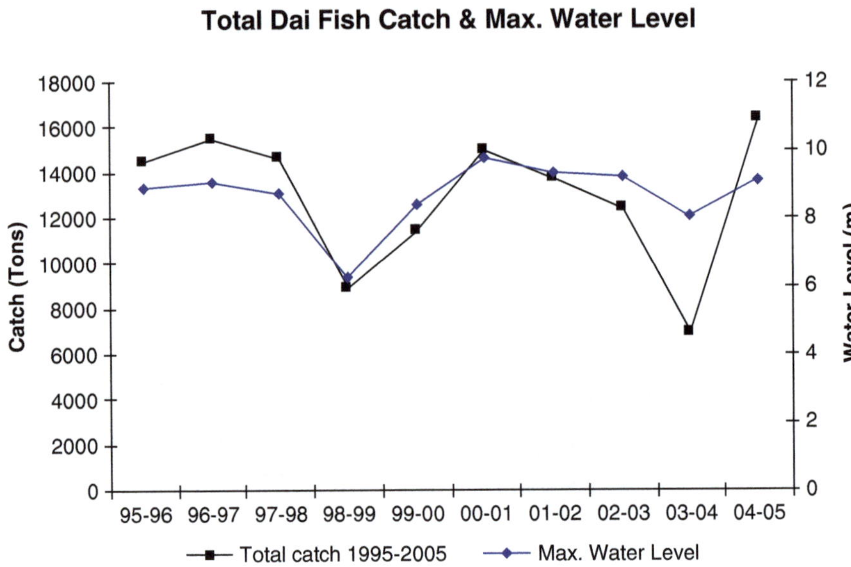

Fig. 2.4 Relationship between total Dai fish catch and maximum water level along the Tonle Sap River (Cambodia)

nutrient regime. Fish migrate to deep pools in the mainstream to take refuge during the dry season, and migrate back to spawning and feeding grounds on floodplains during the flood season. The importance of the flood pulse and morphological dynamism to the river system productivity and sustainability is considered in this study in addition to other key variables for assessing the Mekong River water sector.

Ten year time series data (1995–2005) from the Tonle Sap Lake Dai fishery (dragnet fishery targeting migratory fish species down from Tonle Sap Lake) show a strong correlation between the fish catch, the water level and inundated area (Baran 2005; Catch and Culture 2005). Based on a comparative analysis of the 1995–2005 Dai fish catch coupled with the Mekong River flood levels, Zalinge et al. (2003) maintain that higher floods and associated increase in Tonle Sap flood plain sediment and inundation areas, led to an improvement in survival and growth of fish and fishing yields.[3] The 2003–2004 fish catch of 6,000 metric tons is the lowest since systematic monitoring began in 1994–1995 (Catch and Culture 2005). 2003–2004 was also a period of reduced flow, inundated area and a shorter duration of inundation and increased fishing pressure. The increased Dai fishing catches of 2004–2005 (16,000 metric tons – the highest over the past 10 years) were associated with above average flood levels, longer flood peak duration and reduced illegal fishing (Catch and Culture 2005). The 2003–2005 fish catch and water level data illustrated in Fig. 2.4 point to a sensitivity of fish productivity to Tonle Sap flood levels

[3]ICEM report (2010) estimated that the Mekong marine fish catch is about 0.5 million tonnes/year and reliant on the Mekong marine plume.

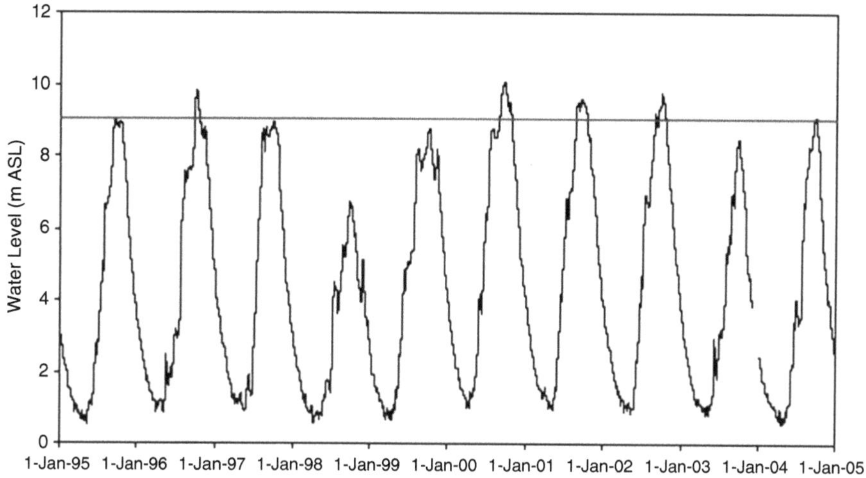

Fig. 2.5 Daily water levels at Phnom Penh port (Tonle Sap River) 1995–2005 (Source: Catch and Culture 2005)

and the spatial extent and duration of flood plain inundation (Catch and Culture 2005). Figure 2.5 illustrates the associated January water levels at the confluence of the Tonle Sap and Mekong rivers for the same period.

Elevation and duration of flood play a key role in the general ecology driving the fish production, as most of the impacts are related to time, height and duration of flood.

2.3 *Observed Long Term Flow Variability*

A time series hydrological analysis at selected assessment points along the Lower Mekong River (annual maximum and mean annual flows from 1924 to 2002) indicates a long term decrease in annual flow at all stations, except for Luang Prabang (Fig. 2.6). The flow decreases point either to a change in the hydrologic response of the major tributary dams downstream of Vientiane or to a decrease in the regional rainfall since the late 1960s (see also Halcrow Group 2003).

The available rainfall data from key stations shows different rainfall trends in different parts of the Mekong basin, with a general decrease in the upper parts of the Basin, and a general modest rainfall increase in the Central Highlands and Mekong delta in Vietnam.

Countries in the lower Mekong Basin are among the most vulnerable to climate change in the world. It is yet to be ascertained how rising temperatures, greater variability of rainfall, rising sea levels, and coastal inundation will affect the basin. According to the Intergovernmental Panel on Climate Change (IPCC),

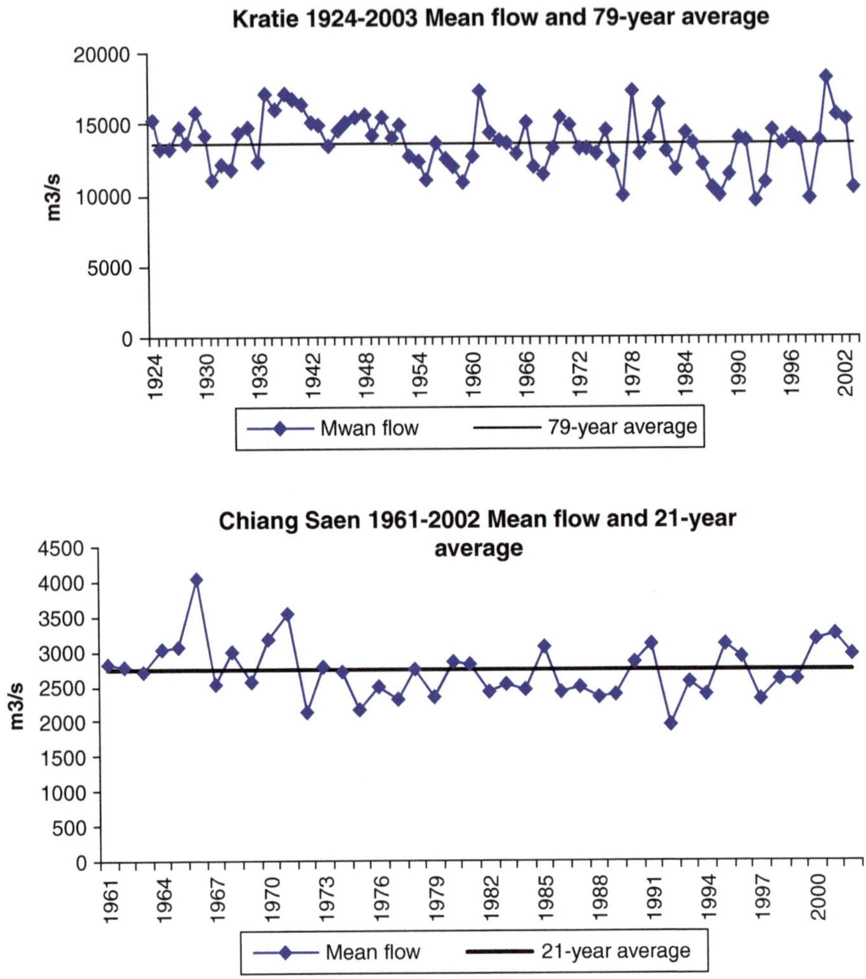

Fig. 2.6 Temporal characteristics of the annual flows at key hydrological stations on Mekong mainstream

the region is likely to experience the upper extremes of the climate scenarios forecast. The preliminary climate change downscaling method conducted by the Environment Division of the MRC predicted that that the mean annual average temperature will increase 0.9°C, 0.7°C, and 0.7°C for the upper Mekong Basin (China) (UMB), lower Mekong Basin (LMB) and the entire Mekong basin respectively. The highest temperature increase is expected in the uppermost part of the Mekong basin. The increase will be less in the LMB but slightly higher in the lower part of the LMB and the delta (MRC 2010d).

2.4 *Water Dependency and Uses in Mekong Region*

The majority of the approximately 278 million people in the wider Mekong Region live in rural areas where they lead subsistence or semi-subsistence agricultural lifestyles (ADB 2009; MRC 2010a). Several millions of people live within a 15 km corridor on the two sides of the mainstream Mekong River and the Tonle Sap River and Lake (MRC 2010b; ICEM 2010).

China's interests in the Mekong Region are focused on developing China's western landlocked provinces to satisfy the energy, water and resource demands of neighbouring provinces (Dore and Yu 2004); and to reduce the rate of emigration to coastal cities (Dosch and Hensengerth 2005). In parallel to hydropower development (a series of 15 dams along the stretch of upper Mekong Region have been constructed or planned Xu and Moller 2003)[4] China is developing water-borne transport infrastructure enabling manufactured exports and raw material imports to and from lower Mekong countries (Wu, Jiabao 2005).

Viet Nam and Thailand are the most intensive water users in the Mekong Region. Thailand has the largest density of roads and hydropower/irrigation projects. Thailand's annual freshwater withdrawals represent almost 30 % of the available water resource stock (ADB and SEI 2002). Ninety percent of Lao PDR and 86 % of Cambodia are located within the Mekong basin respectively, with commensurate proportions of their population dependent on access to the Mekong River Basin resources. Several plans are in place to turn Lao PDR into the "Kuwait" of South East Asia for energy/electricity export (MRC 2005b).

Irrigated agriculture in the Mekong basin is the largest consumptive water user, responsible for 78–94 % of fresh water withdrawals (ADB 2009; MRC 2010a). In 2000, irrigated agriculture use accounted for approximately 15 % (72,837.66 MCM) of the annual average discharge (475,014 MCM), or 80–90 % of the total water abstraction from the MRB. Mekong basin water abstraction is comprised of harvested receding flood water storage, diversion of water from stream and from ground water sources, and water from precipitation (soil moisture) (MRC 2003; MRC 2010b). More than 50 % of irrigated water use occurs in the Mekong delta (MRC 2010c). Estimates of sectoral water withdrawals (including the industry sector) in the Mekong Region are presented in Table 2.3 below.

Water extraction for irrigated agriculture is estimated to continue to grow as most of the MRB remains an agriculturally dominated economy (MRC BDP (Basin Development Plan) 2003), although the use of water for domestic purpose and industrial use is expected to increase to 4.3–5.3 % of available water sources respectively (Papademetriou 2000).

[4]China Hydropower Engineering Association: by 2020 the Lancang River reach will produce a total of 22–25.6 million KW installed capacity from a series of 15 hydropower dams. A cascade of eight hydroelectric power dams with 9.4 million KW installed capacity is being set up in the upper reaches. Another seven dams with 16 million KW installed capacity are built or planned for in the middle and lower reaches of Lancang.

Table 2.3 Water withdrawals in the Lower Mekong Basin, 2000

Country	Total (million m³)	Withdrawals (m³ per person)	Sectoral withdrawals (% of total)		
			Agriculture	Industry	Domestic
Cambodia	4,091	311	98	1	2
China	630,289	494	68	26	7
Lao PDR	2,993	567	90	6	4
Myanmar	33,224	699	98	1	1
Thailand	87,065	1,429	95	2	2
Vietnam	71,392	914	68	24	8

(FAO 2005) Note: Data presented above are for the whole countries

It is projected that by 2025, irrigated agriculture water use will account for about 22 % (104, 503 MCM) of the average annual discharge of the Mekong River and 25–30 % by 2050 depending on the irrigation scale and intensity (Pech and Sunada 2008, MRC BDP 2003). The total irrigation demand in the MRB will be lower than estimated average annual river flow, but this neglects uneven distribution of flow in time and space – severe flow fluctuations between the wet and dry seasons, from wet years compared to dry years, and geographic differences (Pech and Sunada 2008; MRC BDP 2003). In addition, a specified proportion of seasonal runoff is required to maintain environmental flows, aesthetic/recreational services and dependent ecosystems (Ravenga and Mock 2000).

Raskin and Kemp-Benedict (2002) estimate that by the year 2032 the share of domestic and industrial uses will constitute about 20–22 % and 28–29 % respectively of the total water withdrawal in Southeast Asia. Similar projections have been made for the MRB – the domestic and industrial water consumption for 2000 in the Mekong basin was estimated at 2,773.58 MCM or less than 1 % of the average annual Mekong flow. The 2050 domestic and industrial water demand is projected to increase to about 11.5–15.5 % of the total average annual Mekong flow (Pech and Sunada 2003; MRC 2003). Even though the current demand by domestic and industrial water uses remains modest, aggregate water demands for agriculture, domestic and industrial use in 2050 are estimated at between 32 % and 50 % of the total annual flow. This will further increase competition for water resources during the low flow conditions of the dry season and driest years. The increase in domestic and industrial water use leads to a proportional increase in the demand for waste water systems/facilities and improved sanitation that poorly developed, requiring substantial investment and management in many Mekong countries, except for Thailand (MRC 2003).

3 Analytical Approach and Indicators

Traditional analytical approaches have relied on a strong hydrological focus, emphasising practices to control quantity, quality, and timing of water flows. In contrast, the assessment detailed in the following Section focuses on both the availability of water for human use through time and space and the quality and quantity of water required by an aquatic for the to protect and maintain aquatic ecosystem structure, function and dependent species (Smakhtin et al. 2003). Since water is a key strategic resource, vital for sustaining life, promoting development and maintaining the environment, the water sector assessment approach is concerned with both water issues and other closely associated water resources/elements and dependent communities.

A fundamental point of departure deploys a broader perspective that highlights likely changes/impacts resulting from implementation of the six development investments and how those changes are likely to affect interrelated natural resource elements and people's livelihood. Likely changes are particularly important and critical among the rural poor who depend on subsistence livelihoods. A broader set of key indicators and variables to assess the potential water sector changes reflects this broader perspective. The main indicator classes listed in Table 2.4 are: hydrological changes and changes in other water elements; geo-morphological changes that have implication on ecosystem sustainability; and alterations to sustainable livelihoods.

The following sections discuss first the trajectory change induced by implementation of the investment decisions considered in isolation and second the cumulative impacts from a transboundary perspective.

4 Single Factor Impact Assessment

4.1 Hydropower Dams

The magnitude of Mekong River Basin change is function of the number, size and locations of constructed hydropower dams. Dams design, construction and operation also influence the magnitude and valence of change.

4.1.1 Potential Impacts of Hydropower Development

By 2030, the dams in the Mekong tributaries are estimated to substantially alter mainstream river flows (MRC 2010b). For the very first time since river monitoring commenced in 1915, the development sector will alter the hydrological regime of the entire MRB. The active storage will potentially increase by 700 % from 9.9 to 69 km^3. Approximately 23.7 km^3 or 36 % will be located within Yunnan Province, mainly from the two largest hydropower dams with an active storage of 22.2 km^3 (Xiawan and Nuozhadu) (MRC 2010e). The proposed 11 LMB mainstream projects will convert 55 % of the total length of the mainstream stretch between Chiang Saen (Thailand) and Kratie (Cambodia) to reservoir storage. The Lower Mekong River will be fundamentally transformed from a free-flowing river to a series of variable

Table 2.4 Indicators used to assess water sector impacts

Aspects	Indicators/Variables	Triggers
Hydrological changes	Water level	• Substantial impact of climate change is expected only from 2030 and beyond
	Flow level in dry season at specific locations	• Impact of large regional scale forest cover and habitat changes on transboundary water is not critical for an obvious reason
	Flood timing	• Railway link's transboundary impact on water is less obvious
	Flood duration	**Key triggers considered here are**:
	Flood area	• Multi-sector demand and associated intervention(dam, diversion etc.)
	Reverse flow/water level in Tonle Sap;	
	Inundated area, duration and timing	• Impact of large scale mining and processing;
	Storage	• Increase in water demand versus water availability
	Change in water quality (turbidity and relevant quality parameters)	
	Change in salinity intrusion – extent, duration, and concentration	
Other water elements – geo-morphological changes	Barrier effects and dis-connectivity	
	Sediment and nutrient loads from UMB	
	Reservoir conditions	
	Downstream and critical deep pool habitat	
	The state of the Mekong delta	
	Coastal zones	

energy, managed impoundments, characterised by alternate periods of slow water movement and rapidly changing flow in response to dam operations (ICEM 2010).[5] The reduced flood season flows are estimated to reduce the extent and duration of floodplains inundation and contribute to bank erosion on the critical stretches and infilling of deep pools (MRC 2009).

The operation of the mainstream dams can cause significant downstream fluctuations during any 1 day if they are operated as peaking projects. In this case,

[5] Sarkkula et al. 2010. Power Point Presentation at Regional Consultation Workshop of June 2010: About 66 % of the total 1,760 km river distance, Sambor dam (Kratie) site to the upper end of Pak Beng reservoir, will be affected.

water level fluctuations locally may amount to typically 2–4 m or more in extreme cases (MRC 2010c).[6] This may have severe implication for local navigation and the productivity of river bank gardening.

The increase in dry season flow is estimated to be able to meet a planned increase in irrigation abstractions over current levels depending on location along the mainstream. Dams located in the Yunnan reaches of the Mekong River cause a dramatic increase in mean minimum annual dry season flows at all stations, except Kratie. Estimated dry season flows increase by 70 % at upstream stations, reducing to 10 % at the Mekong delta, where notably dry season irrigation demand is highest. About 54 % of the riverbank gardens along the mainstream stretches from Chiang Sean to Kratie will be flooded due to higher dry season flows and reservoir inundation (Sarkkula et al. 2010).

The increases in dry season flow are predicted to increase consumptive water supply and lead to an expansion of commercial irrigated agriculture and land consolidation. Benefits are likely to be accrued by private interests, however these are potentially offset by the incurred public costs associated with the increased use of chemical fertilizer required to offset losses in sediments and nutrients. The variable production costs (water fee, inputs and labour) of agricultural expansion are also predicted to increase.

The change in water surface areas will cause a very significant seasonal redistribution of flow from the wet season to the dry season and reduce sediment transport in the Mekong mainstream, especially to the area above Vientiane (MRC 2010c). The reduction in the maximum water level and increase low flows will be observed over the next decades, associated with increased water storage in large capacity reservoirs. As a result the overall hydrograph will be smoother (Fig. 2.7), especially in the transition from the dry to the wet season (ICEM 2010).

The onset of the transition from dry to wet season will be significantly reduced: 7–8 weeks earlier in Chiang Saen,[7] 2–4 weeks earlier upstream of Pakse and 1 week at Kratie. This change will see a reduction in the important freshwater '*spates*' which drive many ecosystem functions such as fish spawning, lave dripping and fish migration.

There has been an ongoing and active debate about the role of dam reservoirs in regulating flood peak flow ("downstream flood benefit"). Literature insights indicate that mainstream dams in the LMB would only provide limited flood protection to reaches immediately downstream of the reservoir impondment (MRC 2010b, c, d). The peak daily flows will be reduced by 18 % (−1,100 m^3/s) at Chiang Sean, 15 %

[6] Sarkkula et al. 2010. Personal communication: The time elapse for a rapid fluctuation from opening the turbines – planned and unplanned circumstances, and breakdowns of plant and electrical transmission systems at the proposed Xayaboury dam site, will be about 1–1 ½h to the city of Luang Prabang, very little warning time for bank-side residents to prepare for inundation.

[7] MRC 2010b: At Chiang Sean, peak daily flows will be reduced by 18 % (1,100 m^3/s), and dry season flow volume will be increased by about 61 % (12,093 MCM), and peak daily flows at Kratie will be reduced by 7 %, and dry season volume will be increased by 23 %.

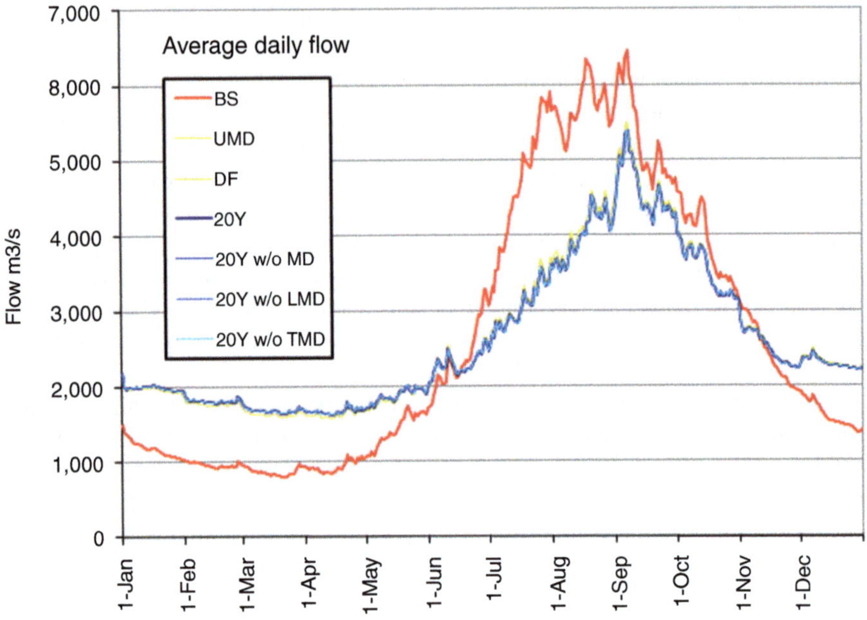

Fig. 2.7 Changes in hydrology at Chiang Sean; BS – Baseline Scenario…; UMD – Upper Mainstream Dam Scenario.; DF – Definite Future Scenario:…; MD – Mainstream Dam:…; LMD – Lower Mekong Mainstream Dam:…; TMD – Thai Mainstream Dam:…

(–2,381 m³/s) at Vientiane, 7 % (–3,456 m³/s) at Kratie and only 4 % (–855 m³/s) in the Viet Nam Mekong Delta (Tan Chau). Over estimates of the flood reduction functions of dams may lead to a false sense of security in the face of historical floods (100–500 year recurrence flood), as will dam failure due to earthquakes and uncoordinated, abrupt reservoir releases of flooding waters. Records from around the world indicate that flood protection (which does not bring in revenue) tends to be neglected in multipurpose projects (Regional POE 2010).[8]

[8]Most of the dams have not been designed to take into consideration major natural disasters such as earthquakes or floods. Lessons from the typhoon Ketsana (end of September 2009) show that mismanagement can inflict significant losses on vulnerable communities. Even where they have been considered, the imperatives of maximizing revenue obliges dam operators to maintain maximum reservoir levels (and thus maximize hydropower generation as well as irrigation water availability). In recent years a number of earthquakes occurred across Asia resulting in impacts to numerous dams (Mongabay.com 2008; Brewer 2008; Vijay and Ramesh 2005; Hough and Martin 2001). For example, the 12 May 2008 earthquake in Sichuan province of China (7.9 magnitude) seriously damaged hydroelectric dams and caused serious social and economic losses. Sixty-nine dams were in danger of collapse, 310 were at "high risk," and 1,424 posed a "moderate risk" (Brewer 2008). China said it would spend more than $1.3 billion per year fixing vulnerable dams, many of which were poorly constructed (Mongabay.com 2008; Brewer 2008). USGS Earth Quake Hazards Project reported two major quake measuring 4.7 on the Richter scale in late February 2011 and 6.1-magnitude quake in 2007at the proposed mainstream dams near Xayaboury, Lao PDR.

There will be an overall 7 % reduction in flooded area (309,000 ha) in an average year. The reduction areas are expected to be smaller in wet years and larger in dry years. The greatest area of reduction occurs in Cambodia (142,000 ha), Lao and Thailand (17 % and 19 % reduction respectively) (MRC 2010d; MRC 2009)

Estimates for 2030 indicate the dam development scenarios will result in significant changes in the ecology of the Tonle Sap Lake. The inundation of the Tonle Sap Lake will be reduced by 5–10 % (500–600 km^2), the reverse flow in Tonle Sap will start at least a week sooner, and the average days of reverse flow will be shorter (about 8 days). Dry season inundated area is anticipated to increase by 5–8 % converting a seasonal terrestrial ecosystem into permanent aquatic. These changes will affect ecosystem and farming productivity, fish migration and sediment flushing capacity. The decrease in reverse flow volume to the Tonle Sap Lake will result in a reduction of flooded area, flood depth and duration; and a reduction in sediment inflow into the lake and blockage of fish migration paths by mainstream dams (ICEM 2010).

With mainstream hydropower projects operating, there will be much less velocity to suspend particles and keep them moving, and the result of this will be enhanced sedimentation, with the formation of deltaic type deposits at the head of each of the reservoirs, and middle and lower parts of each reservoir associated with reduced velocities/gradients (MRC 2010d; ICEM 2010). Only the loads of suspended (fine sized) particles has been measured at several stations on the Mekong mainstream since the 1960s. The development will also potentially reduce fined sized sediment transport from 70 % to 80 % (75–81 % reduction in sediment load) from the UMB (from 90 to 20 Mt/year at Chiang Sean, and from 165 to 88 Mt/year at Kratie) (Sarkkula et al. 2010). The reduction will result in significant nutrient losses in floodplains and coastal regions. Sediment reductions will reduce the productivity of farming and fisheries within and beyond the Mekong. Long-term changes to river bed and bank erosion rates are likely, including increased threats to the stability of the delta shaping processes, potentially exacerbated by sea level rise.[9]

The reservoirs fisheries are expected to increase fish captures by 10,000–30,000 t or US$ 40 million (MRC 2010b), but the Mekong capture fishery of 2.1 million tons or 22 % of world fresh water fisheries, will be at risk. The 2030 scenario without Mekong mainstream dams will reduce the Mekong fishery by 210,000–560,000 t/year (10–27 % reduction) and 12 mainstream dams will cause direct loss of another 340,000 t/year (US$476 million) or another 17 % of total catch. Aggregate reservoir fisheries could compensate up to 10 % of the lost capture fish production predicted to occur in the absence of mainstream dams. It is not clear about the magnitude and time scales that change will have impact on the Mekong marine fishery (currently estimated at 500,000 t/year or a replacement value of about US$ 40 million). Reductions in boat and fishing tackle manufacture, salt and

[9]Loss of sand-sized sediments to Mekong Delta and marine environment result in loss of nutrients (phosphates) to agriculture = 3,400 tonnes/year = US$ 24 million in replacement value/year, and reduction in nutrient loads to over 18,000 km^2 of Cambodia flood plain and 5,000–10,000 km^2 of Mekong Delta floodplain and Mekong marine sediment plume

ice production, and fish processing estimated at US$ 2–4 billion (MRC 2010c) are predicted as consequences of fishery losses.

Navigability will be substantially improved with lower requirements for channel improvement to provide cheaper and more affordable means of transport. However, with improvement in road and rail link, river navigation is not seen as a high priority. The 2030 development scenario will also introduce additional barriers and increase river disconnection as 37–81.3 % of the watershed will be obstructed. Table 2.5 summarises the estimated effects of hydropower development on the Mekong water sector.

4.2 *Water Diversion*

Possible water diversions/extractions from the main stream of the river system are generally expressed as the amount of water extraction at various places required for agriculture and other consumptive uses. Several combinations of inter-basin diversions from the Mekong River and tributaries out of the basin, and intra basin diversions – from the Mekong River and its tributaries into another part of the Mekong basin have been either proposed or subject to member country consideration (MRC 2005). It is assumed that the principle idea of diverting water from a Mekong tributary located on Lao PDR, i.e. diversions from the Nam Ngum into Isaan, Thailand is likely to occur within the duration of this assessment (2010–2030). The study assesses the implications for water availability during the dry season in the relevant catchments and downstream locations from the outtake. Lao PDR, Cambodia and Vietnam are the downstream countries considered.

Intra-basin diversions for irrigation in north-east Thailand, includes diversion of water from the Lao PDR tributary (Fig. 2.8). Based on the consultancy report by Sanyu Consultants Inc (2004), the Nam Ngum Water Management Project envisages both an irrigated agriculture development in the Vientiane Plain, and "surplus" water diverted into northeastern Thailand. Water will be diverted to Issan via a Nam Ngum barrage/diversion dyke located about 6 km from the Mekong confluence, creating a hydraulic head sufficient for gravity diversion into a siphon tunnel under the Mekong (Sanyu 2004; Molle and Floch 2008). A constant diversion rate of 300 m^3/s for the whole year has been proposed, apportioned equally to the Thailand headwaters of the Mun River and Chi River (or a total of 6,878 MCM/year)

4.2.1 Potential Impacts from the Proposed Nam Ngum to Northeast Thailand Diversion

Initial qualitative assessments indicate the proposed diversion cumulatively contributes to reductions in dry season flows, detected primarily from December to March

Table 2.5 Summary of impacts on water sector from hydropower development

Aspects	Indicators/Variables	Changes
Hydrological changes	Water level	The development sector is of sufficient magnitude to alter the hydrological regime of the entire MRB (Active storage increases from 9.9 to 69.8 km^2)
	Flow level	Dry season flow increases by 70 % at upstream stations, and 10 % at Mekong delta
	Flood timing	Reduction in the onset of transition from dry to flood season – 7–8 weeks earlier in Ch. Saen, 2–4 weeks upstream of Pakse, 1 week at Kratie
	Flood duration	To significantly affect uppermost of LMB (above Pakse and Vientiane)
	Flood level and area	Flow decrease by 18 % at upstream stations, but only 2 % at delta. Typical reduction in flooded area
	Reverse flow/water in Tonle Sap	Reduction in duration and volume of reverse flow
	Inundated area, duration, timing	Dry season inundated area increase significantly
	Storage	700 % times increase in active storage area
	Change in water quality (turbidity and relevant quality parameters)	Reduce sedimentation transport from 70 % to 80 %. Loss of nutrients in floodplains and coastal offshore
	Change in salinity intrusion – extent, duration, concentration	Increase in 5–10 % increase in flow, but combined with increased water diversion and climate change, salinity extent further increases
Other water elements – geo-morphological changes	Barrier and dis-connectivity	From 37 % to 81.3 % of the watershed will be obstructed
	Sediment and nutrient	Reduction in-stream power – lost energy in moving over the riverbed and turbulent flow dissipation. 75–81 % reduction in sediment load
	Changes in reservoirs	Siltation and formation of deltalic areas,
	Downstream and critical deep pool and habitat;	Downstream erosion and long-term impact on 48–70 % of deep pools habitat. Loss of natural fertilizer and food-chains
	Mekong delta	Erosion, instability of shore-lines – land mass advances by 60–80 annually comes to an end, at location annual erosion up to 45 m/year
	Coastal zones	Loss of nutrient and impacts on marine fisheries, but magnitude and time-scales unclear

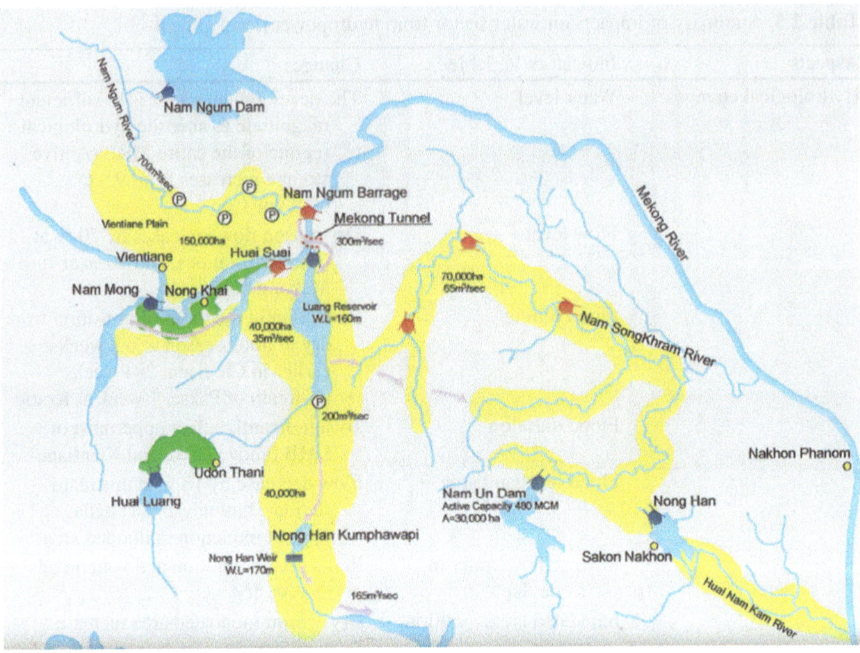

Fig. 2.8 Proposed diversion site map – Nam Ngum Water Management Project for Vientiane plain of Lao PDR and Northeast Thailand (Source: Sanyu 2004)

(MRC 2005). Additional flows from upstream reservoir operation may compensate for the reduced flows. The combined influence of proposed inter-catchment water diversions and upstream dam operations introduces the potential for both positive and negative transboundary outcomes for this stretch of the Mekong River (the Thalweg channel) that Lao PDR and Thailand share as their political boundary. Positive outcomes for Thailand relate to additional irrigation water, however Sanyu (2004) flags the potential flood hazard of reservoir operations in the wet season. For example the Huai Luang Reservoir will need to be completely emptied by the end of July in order to store the August flood waters emanating from the Huai Luang basin.

Intra-basin diversions can cause significant reductions in dry season flows in the Nam Ngum/Nam Lik tributaries as sources of diverted water. Abstractions are likely to cause more severe impacts on mainstream reaches by-passed by the diversion, especially in drier years. A constant annual diversion rate of 300 m^3/s from the Nam Ngum may potentially cause water shortages during critical dry season months, since the observed average flow (1993–2000) at Ban Pak Kannhoung (further upstream from the diversion point), was between 791 and 686 m^3/s from January to April (Sanyu 2004). The abstraction of nearly 50 % of dry season flow assigned for diversion to north-east Thailand will cause water shortages for other uses including the projected abstraction for irrigation in the Vientiane Plain.

Approximately 250 m^3/s is projected for irrigating 150,000 ha in the north and south Vientiane Plains, and at least 570 m^3/s for maintaining minimum flow in the Mekong mainstream (Sanyu 2004). However, it is expected that the planned release of controlled flows from Nam Ngum Dams 1, 2, and 3 and flows from the Nam Link development may help relieve that water shortage during critical dry season, assuming reservoir releases and operations are coordinated.

The reduction of dry season flow without compensation from other flow coupling with climate changes may potentially deteriorate salinity intrusion in Mekong delta. Table 2.6 summarises the estimated impacts of a Nam Ngum water diversion to northeast Thailand on the Mekong Water sector.

4.3 Large-Scale Mining Activities

Bauxite is one of the most economically important minerals. According to the U.S. Geological Survey, Asia contains about 17 % of the total global bauxite reserves (ranked 3rd after South America and Africa). Cambodia, Lao PDR and Vietnam are reportedly rich in mineral resources; however, the exploitation of these resources has typically been delayed due to conflict, lack of foreign investment, and limited capital and capacity to establish extensive mining operations (Lazarus 2009).

Deposits of bauxite have been prospected and explored for the last 5 years in the Bolaven plateau in southern Lao PDR (Champassak Province next to the Cambodian border, see Figure 5 Chap. 1). A commercially viable alumina refinery consumes electricity in the order of 600–800 MW, dependent on the technology applied and the scale of the operation, and large quantities of water (Sekong River and Sekong dams) (Lazarus 2009). A large concession area would be required and would have to be fully cleared along the Bolaven plateau.

Substantial water sector impacts are expected as a result of large-scale bauxite mining in the Bolaven plateau, including altered water quality, transport of sediments, soil erosion, land use change and loss of forest cover (Lazarus 2009, ADB – RETA 40082 2011). Land use change and loss of forest cover resulting from forest clearance and earth removal for mining (open cut mines, or removing the surface layers) can potentially cause significant soil erosion, sedimentation and altered runoff characteristics. Deforested areas, especially those having steep slopes and fragile soils change runoff conditions and will potentially generate deteriorated flood conditions, including sudden flash floods, and more intense drought (ADB – RETA 40082 2011).

Changes to sedimentation loads and soil erosion will affect the Mekong and Sekong Rivers although lagoons and tailing ponds may be used to reduce the amount of sediment actually reaching the rivers (Lazarus 2009, ADB – RETA 40082 2011).

It is also expected that large quantities of water will be extracted either from the river or as groundwater for the extraction and production of minerals, e.g. for washing of ores and in the production processes (Lazarus 2009, ADB – RETA 40082

Table 2.6 Summary of potential impacts from Nam Ngum water diversion

Aspects	Indicators/Variables	Changes
Hydrological changes	Water level	Intra-basin diversions can cause significant reductions in dry season flows in the mainstream reaches by-passed by the diversion. Tributaries from where water is abstracted will be more severely impacted. Potential flood hazard due to storage reservoir operation in north-east Thailand
	Flow level in dry season at specific locations	Short term water shortage for contributing to maintaining minimum flow in the Mekong, and meeting water demands in Vientiane Plain in critical months of dry seasons
	Flood timing	Flood timing is affected by dam regulation from upper catchment
	Flood duration	
	Flood area	
	Reverse flow/water level in Tonle Sap;	
	Inundated area, duration and timing	
	Storage	Series of 300 new large and medium-sized reservoirs and 25,000 community reservoirs in north-east Thailand
	Change in water quality (turbidity and relevant quality parameters)	Point-source and non-point-source pollutants are expected to substantially increase for irrigation expansion and human activities in Vientiane Plain and from return flow from Mun Chi (Thailand)
	Change in salinity intrusion – extent, duration, and concentration	May significantly contribute to worsening salinity intrusion in Mekong delta
Other water elements – geo-morphological changes	Barrier effects and disconnectivity	Habitat partitioning (locally) from water manager devices (WMD)
	Sediment and nutrients	Sediment trapping is caused by series of 4–5 dams on Nam Ngum and Nam Lik
	Changes in reservoirs	Ibid
	Downstream and critical deep pool and habitat	Some noticeable impact on landscape and habitat from diversion channel or siphon structure construction
	Mekong delta	No significant impact
	Coastal zones	No significant impact

2011). ADB – RETA 40082 (2011) estimated that one of the bauxite mines on the Bolaven plateau is expected to use about 110,000 m^3/per day, sourced from the Xe Namnoy River, equivalent to half of the river's peak flow. The local and transboundary low flow issues are likely to be amplified during the dry season and during drier years for downstream communities in both Lao PDR and Cambodia (ADB – RETA 40082 2011).

Pollution derived from mining activities affects mainly surface water and potentially groundwater. Toxic compounds and metal salts are likely to leach into extraction waters, concentrating in sediments and eventually discharged into streams and rivers. The generation of highly alkaline red mud and associated contaminants represents the most significant risk to downstream surface and groundwater sources. Poorly managed disposal will lead to deterioration of livelihoods of dependent communities in Lao PDR and Cambodia, declining public health and a decline of downstream fish stocks and aquatic life (ADB – RETA 40082 2011).

Large scale mining will trigger otheris dependent on the development of major infrastructure such as railway, road transport and energy generation. The production of aluminium requires a great deal of energy. The mining of bauxite and production of alumina require about 200–250 MW for 1 t of alumina, and a modern aluminium smelter would require about 14,000 MW to smelt 1 t of aluminium. Cheap energy is essential for cost effectiveness in aluminium production, estimated at 2.5–3.5 US cents per kilowatt. Lao PDR is currently selling hydropower generated electricity to Vietnam and Thailand at 5–6 US cents per kilowatt. At present this makes localised aluminium smelting unviable and the transporting of locally smelted alumina elsewhere for the production of aluminium a likely production strategy.

Transport infrastructure will need to be further improved through development of an east–west road corridor to connect Thailand through Lao PDR to Vietnam and north–south road corridor to connect southern China (Kunming) through Lao PDR to Thailand (Thompson Richard 2010). Table 2.7 summarise the effects of bauxite mining on the Mekong water sector.

4.4 Mekong Rail Link Within Wider Mekong Region Economic Corridor

Substantial transport infrastructure developments have been proposed for the wider Mekong Region, introducing substantial improvements in regional trading opportunities (ADB 2009). In contrast, there exist countervailing concerns regarding the construction and operation of connecting rail and road networks. Large scale transport infrastructure has the potential to introduce additional pressures on terrestrial and aquatic ecosystems and altered patterns of migration with associated social pressures.

Table 2.7 Summary of potential impacts from large-scale bauxite mining on Bolaven Plateau

Aspects	Indicators/Variables	Changes
Hydrological changes	Water level	Large quantities of water will be extracted either from the river or ground water for mining operations in the extraction and production of minerals
	Flow level in dry season at specific locations	Local and transboundary low flow issues can be escalated in dry season and drier years for the communities living downstream in Lao PDR and Cambodia
	Flood timing	
	Flood duration	Cause significant increase in local run-off
	Flood area	Deforested areas, especially those having steep slopes and fragile soils increase the runoff conditions and potentially generate stronger flood, including sudden flash floods, and stronger drought events
	Reverse flow/water level in Tonle Sap;	
	Land use changes	Large land area and forest cover will be cleared for mining, transport facility, smelter facility, and energy production and transmission line
	Storage	
	Change in water quality (turbidity and relevant quality parameters)	Pollution of surface water and potentially ground waters from: • Toxic compounds, metal salts etc. into the water and hence discharged into streams and rivers • Build up of toxic materials in the sediments. Generation of highly alkaline red mud
	Change in salinity intrusion – extent, duration, and concentration	
Other water elements – geo-morphological changes	Barrier effects and disconnectivity	Significantly large area of landscape will be disturbed
	Sediment and nutrients	Cause significant soil erosion, sedimentation and run-off changes -discharge of sediments into water courses (Mekong River and Sekong River etc.)
	Changes in reservoirs	
	Downstream and critical deep pool and habitat	Fish and wildlife habitats will be disturbed or affected by pollution from mining
	Mekong delta	
	Coastal zones	

In August 2010, Ministers from Cambodia, China, Lao PDR, Myanmar, Thailand and Vietnam adopted the plan which they called "a significant first step toward the development of an integrated railway system" (ADB 2009). The existing six Nations' national railway systems do not connect except for a line that connects China and Vietnam. Lao PDR has no existing rail network. The rail network plan coincides with an effort by Mekong Nations to develop "economic corridors" based on new road linkages. The most viable of four possible routes cites Route 1 – which would link Bangkok to Phnom Penh, then Ho Chi Minh City and Hanoi, and finally up to Nanning and Kunming, largely using existing lines or those already under construction (ADB 2009). This Section is focused on large scale transport infrastructure, in particular the connected railway project.

By 2025, the regional railway lines in all countries will be developed (or rehabilitated) to further regional integration and connectivity in the Mekong Region.[10] (see Figure 4 Chap. 1)

Built infrastructures consist of a variety of man-made structures that contribute to modified natural and social systems. For instance, dams, weirs, irrigation schemes, and dykes can potentially alter water outflow, whilst embankments, polders, levees and roads prevent the natural exchange and movement of water fish, sediments and nutrients. Since major infrastructures are usually constructed to enhance socio-economic development, they tend to attract people and industry: therefore, river and terrestrial ecosystems containing these structures must also contend with the resulting environmental and social pressures that are independent from, and in addition to, the direct influences of built infrastructure on the physical and biological dimensions of the system (Baran et al. 2007).

The key impact will be alteration of surface hydrology due to road crossing, resulting in modified sediment deposition and increased soil erosion, and modification of water local flood patterns. The impacts can potentially be trans-national where the structures intersect key trans-boundary ecosystems and communities. Table 2.8 summarises the estimated impacts of proposed railway networks in the wider Mekong Region on the Mekong water sector.

The modification in water flows and flood patterns are the predominate threats to the ecology of floodplains by poorly designed and built roads and rail-links. The study by a Finnish group and the World Fish Centre analysed numerous experiences

[10]On going and planned projects include:

- Cambodia: rehabilitation of the railway funded by ADB and AusAID;
- China is financing the Feasibility Study for line between Phnom Penh and VN Border;
- China is building new line to the VN and Myanmar borders
- Thailand's high-speed train line and new links to link Lao and onward to VN;
- VN considering lines to Lao and Cambodia; and
- Railway through Lao PDR will be online in 2015 and form part of the Asian-China railway, which runs from Yunnan Province south through Lao PDR to Thailand, Malaysia and Singapore.

Table 2.8 Summary of impacts from major railway links

Aspects	Indicators/Variables	Changes
Hydrological changes	Water level	Alteration of surface hydrology crossed by roads, resulting in increased sediment and increased soil erosion mostly locally. Changes in the level of the water table
	Flow level in dry season at specific locations	Increase water levels by 40–50 cm up to 20 km upstream within local floodplain
	Flood timing	
	Flood duration	
	Inundated/Flood area	Modification in local water flows and flood patterns
	Reverse flow/water level in Tonle Sap;	
	Inundated area, duration and timing	
	Storage	
	Change in water quality (turbidity and relevant quality parameters)	Deterioration of surface water quality due to silt runoff and sanitary wastes
	Change in salinity intrusion – extent, duration, and concentration	
Other water elements – geo-morphological changes	Barrier effects and dis-connectivity	Habitat partitioning locally but with transboundary implication at cross border points. Restriction of fish and wildlife migrations (locally)
	Sediment and erosion	Increased local sediment and increased soil erosion
	Changes in reservoirs	
	Downstream and critical deep pool and habitat	
	Mekong delta	
	Coastal zones	

in tropical floodplains in Asia, Africa and South America and reviewed more than 300 journal articles, reports and books. Their review confirms the need to consider and avoid irreversible changes to water when planning infrastructure development. In particular changes affecting seasonal flooding or disrupting the natural "connectivity" between various water bodies have to be avoided when building major infrastructure in critical watershed and floodplains (see e.g. Baran et al. 2007).

4.5 Sea-Level Rise and Adaptation Strategies in Vietnam's Mekong Delta

The important aspect for the assessment is to consider water sector implications of climate change induced sea-level rise for the Vietnamese Mekong Delta. Current climate change forecasts for the Mekong delta identify increased incidents in floods and droughts together with changes to seasonal rainfall patterns and an increased incidence and severity of typhoons and storm surge.

Sea level rises combined with increased water consumption within the Mekong delta and upstream in the dry season and droughts will increase levels of inland saline intrusion.

The Vietnamese government has identified substantial increases in 'hard' water infrastructure investments (higher embankments and strengthening of dikes and roads, sluice gates) as a primary strategy to offset any direct detrimental consequences of sea level rise. Proposed structural measures will cause hydrological and morphological changes such as obstructions to flood drainage, increasing flood peaks, and increasingly localized retention of sediments. The infrastructure strategy may also decrease the amount of overbank flow across the border between Cambodia and Vietnam with some significant transboundary consequences such as back-up flooding waters into Cambodia. Couple with the flood control structures along the Viet Nam and Cambodian Border and Viet Nam Mekong Delta, additional infrastructure to mitigate seavel rise may also lead to decreasing drainage to the West Mekong Region increased drainage to the Gulf of Thailand (Shigeko 2009). Table 2.9 summarises the estimated effects of sea level rises in the Mekong delta on the Mekong water sector.

4.6 Large-Scale Rubber Plantations

It is estimated that 1.6 million hectares of additional rubber will be planted by 2030. According to various resources, there were 2.9 million hectares of rubber planted in the Mekong Region in year 2000/2002. According to Fox (2012), 4 % of current vegetation in 2000 would be replaced by rubber plantation by the year 2050. Ziegler et al. (2009) estimated that more than 500,000 ha of native vegetation may have already been converted to rubber. By 2050, the area of land dedicated to rubber and other diversified farming systems may double or triple.

There is limited literature based insights and research concerned with the affects of large-scale tree plantations on water, flora and fauna of the zones proximate to plantations (Grupo Guayubira 2011). The impacts on local lakes, streams and rivers can be severe due to hydrological modifications, forest cover changes and soil

Table 2.9 Summary of potential impacts of climate change induced sea level rises in the Mekong delta on the Mekong water sector

Aspects	Indicators/Variables	Measurable changes
Hydrological changes	Flow level in dry season at specific locations;	
	• Times – wet and dry seasons, wet, dry or average year;	
	• Reverse flow/water level in Tonle Sap;	
	• Delta	Sea level rising together with sea water intrusion structures obstructing directly flood discharge to sea
	Change in inundated area, duration and timing	Causing flooding in a wider area and longer flooding duration. Duration of the ending flood drainage for the entire delta would be longer
	Change in water quality (turbidity and relevant quality parameters)	Stagnation of flow affects flushing capacity. Further deterioration of water quality in the delta areas of Viet Nam
	Change in salinity intrusion – extent, duration, and concentration	
	Barrier effects and dis-connectivity	Sea water intrusion and flood control infrastructures will cause further ecological and habitat dis-connectivity in Mekong delta
	Change in sediment transport	Trapping of sediment behind dykes and road networks
Geo-morphological changes	Sedimentation and erosion rates	
	Downstream;	
	Critical sub-catchment;	
	Critical deep pool and habitat;	
	Mekong delta	Significant

erosion (Menne 2004; CSIRO 2011). Studies on forestry plantation water use confirm that plantation expansion has had an impact on catchment stream flow. The combination of changes in climate, vegetation, soil, geology and other features can make a difference to water flows. However, further studies are needed to account for variability due to tree growth and management, such as planting date and location within a catchment, which impact on stream flow and sedimentation within the transboundary context.

5 Cumulative Assessment

5.1 *Current Practice of Impact Assessment in the Mekong Region*

Combining identified multiple stresses, risks and various degrees of vulnerabilities in an assessment can provide critical insights for regional planning and development of the Mekong basin water and other key resources. Focused on the water sector this Chapter has attempted to account for both cross-sectoral implications of development interventions and their likely impacts on natural resources and on communities (ADB and SEI 2002; World Bank and ADB 2006).

In addition to the existing rapid pace of habitat and watershed degradation, the development and operation of large-scale hydropower, major reservoirs, increased irrigation and water diversion in different parts of the wider Mekong Region are likely to create substantial positive and negative outcomes for communities and natural resources (Pech 2010).

In spite of some attempts to create a framework for assessing the impact of various development projects in the Mekong River Basin, to date there has been no "cumulative assessment" carried out (Keskinen and Kummu 2010). The absence of such a comprehensive impact assessment and an underpinning monitoring system in the Mekong Region highlights the concern for severe cumulative impacts on resource sustainability and livelihoods.

5.2 *Expert Opinion on Combined Impacts*

Based on expert opinion and the evidence provided in the previous Chapter sections, this Section aims to provide an assessment of the combined impact on the water sector assuming all pending development investments were to be realised in the near future.

All projects and plans for developing hydropower, intensifying water abstraction for irrigation and large scale mining activities are likely to create region wide impacts. The construction of roads and similar structures in the watershed and floodplains, and large scale rubber plantation potentially cause further hydrological and morphological changes but mostly with local implications. The hydropower and diversion of water for agriculture across the Mekong basin are presently the most important economic activities of interest and the focus of most inter sectoral concern.

It is not feasible to assess the impacts of each and every dam and irrigation project due to the fact that they are so numerous and subject to variable data availability and reliability. It is important to note that the current modelled hydrological changes utilised by the MRC to derive conclusions on cumulative outcomes are built on a number of rough assumptions (MRC 2010b). One thing is clear that these major development projects, considered incrementally and cumulatively, will change the hydrological regime of the Mekong River Basin and associated water elements (World Bank and ADB 2006). Table 2.10 summarise the proposed cumulative effects on the Mekong water sector.

5.3 *Hydrological Changes*

5.3.1 Water Level

As shown in Chap. 2, the development sector will alter the hydrological regime of the entire MRB – mostly due to hydropower development, for the first time since monitoring commenced in 1915. The active storage will potentially increase by 700 % from 9.9 to 69 km^3 and about 23.7 km^3 or 36 % will be within Yunnan Province, mainly from its two largest hydropower dams with active storage about 22.2 km^3 (Xiawan and Nuozhadu) (MRC 2010e). With the construction of 11 proposed LMB mainstream projects, 55 % of the total length of the mainstream stretch between Chiang Saen (Thailand) and Kratie (Cambodia) will be converted to reservoir, transforming a "live" river to a series of managed impoundments, typified by slow water movement and highly modified flow patterns in response to dam operations (ICEM 2010)

5.3.2 Flow Level in Dry Season at Specific Locations

The increase in dry season flow will theoretically be able to meet a planned increase in irrigation abstractions over current levels depending on location along the mainstream. However, China and other LMB dams will cause an increase in the lowest and mean minimum annual dry season flows at all stations, except for Kratie where the increase is restricted to approximately 2 % and 10 % respectively (MRC 2010b; Halcrow Group 2003; Halcrow Group 2004). Dry season flow increases are the lowest in the Mekong delta, where, conversely the dry season irrigation demand is the highest. Hence, dams have strongest affect in the areas nearest to the dam sites, not in the Mekong delta where irrigation water demand is highest.

The operation of the mainstream dams can cause significant downstream fluctuations during any 1 day if they are operated as peaking projects. In this case, water level fluctuations locally may amount to typically 2–4 m or more in extreme cases (MRC 2010c) adversely affecting underprivileged farmers relying on the riverbank gardening and fisheries.

Water abstraction from the Mekong River is limited during the wet season when flow levels are high and rain water is available, however, there are many constraints on water utilisation during the dry season, especially in the drier years. The drier years pose the most severe water constraints.

Table 2.10 Summary of potential cumulative impacts

Aspects	Indicators/Variables	Foreseeable 20 years 2030 with LMB dams
Hydrologic changes	Development sector totally alter hydrologic regime of the entire basin	
	Water level	Mainstream flow fluctuation relatively slow +/−0.16 m/day at Luang Prabang, +/−0.11 m/day at Pakse and 0.09 m/day at Stung Treng. But further changes depend on dam operation and regulation (peaking vs. continuous). Areas immediately (40–50 km) below reservoirs can experience up to 3–6 m of daily flow fluctuations from peak operations and abrupt release. Potential flood hazard due to storage reservoir operation in the receiving basin
	Flow level in dry season at specific locations;	Mainstream flow change due to mainstream dams has the least positive effect to support any major water diversions from Tributaries. Tributaries from where water is abstracted will be more severely impacted. Intra-basin diversions can cause significant reductions in dry season flows in the mainstream reaches by-passed by the diversion. 25–50 or 70 % increase in dry season in Northern Lao PDR and Thailand (higher than annual deviation), but only 10 % in delta. Increase in irrigation water use coupling with sea water rising will further exacerbate water shortage in extreme drought period in delta. Large quantities of water will be extracted either from the river or as groundwater for the extraction and production of minerals → local and transboundary low flow issues can be escalated during the dry season and during drier years for downstream communities in Lao PDR and Cambodia
	Flood timing	More even hydrograph in stretch above Vientiane transition from dry to flood: 2–4 weeks shorter above Pakse, and by 1 week in Cambodian floodplains. Spates and first flushes of transition to flood no longer occur. Further reduction the length of transition period to flood season (disappearance totally above Vientiane). Emergency release due to flooding to avoid dam break will cause flooding in the immediate downstream stretches
	Flood duration	Shorter flood peak – Mostly at uppermost reaches Lower and shorter flood peak
	Flood area	Reduction of 300,000 ha in flooded area (15 %) in Thailand and Lao PDR, 5 % in Cambodia and Viet Nam 9000–150,000 ha of garden and agricultural land inundated in dam reservoir sites (54 %) in Chiang Saen to Pakse. Disappearance of floodplains and channels due to impoundment of river stretches into reservoirs. Downstream: reduction in wetland by 34 % in Lao PDR, 18 % in Thailand, 2.4 % in Cambodia

(continued)

Table 2.10 (continued)

Aspects	Indicators/Variables	Foreseeable 20 years 2030 with LMB dams
	Reverse flow/water level in Tonle Sap	Reduce flooded area by 500–600 km^2 (−5 to 10 %) – affect reverse flow in and out. Increase in dry season inundated area +5–8 % Some 82 % of the total flooded area was subject to shorter flood durations under the worst case average annual flow into TLS will be reduced by 13 % or over 430 MCM
	Inundated area, duration and timing	Significant reduction in flooded areas and wetlands in LMB floodplains and Tonle Sap. Sea water intrusion prevention in VN delta causes flooding in a wider area and longer flooding duration. Duration of the ending flood drainage for the entire delta would be longer
	Storage	700 % increase in active to store 14 % of the mean annual flow (9.9–68.8 km^3). Series of 300 new large and medium-sized reservoirs and 25,000 community reservoirs in north-east Thailand as result of water diversion schemes
	Change in water quality (turbidity and relevant quality parameters)	Reduced turbidity by 75 % Overall nutrient loading from irrigated areas increases with 85 % (N) and 100 % (P) Overall nutrient loading from wastewater discharges increases with 33 % (N and P) Considerable basin-wide increase in herbicide and pesticide/fungicide use (75 % and 59 %) Point-source and non-point source pollutants are expected to substantially increase for irrigation expansion and human activities in Vientiane plain and from return flow from Mun Chi (Thailand). Increases are largest in the Mun – Chi Basin, the 3Ss Basin (agricultural run-off and mining operations) and the Tonle Sap Basin Large-scale mining in Bolevan will cause serious water quality issue locally, but there will be long-term trans-boundary water quality issues in 3Ss region. Investment in water infrastructure for salinity intrusion prevention will reduce flushing capacity that lead to further deterioration of water quality in the delta areas of Viet Nam
	Change in salinity intrusion	Increase in dry season flow from dam operations will be offset by sea level rise by 0.3 or 1 m by 2,100
Geomorphologic changes	Barrier effects and dis-connectivity	55–66 % of the total LMR 1,760 km river distance (Sambor to Pak Beng) and 81.3 % of the watershed will be obstructed
		Major changes to sediment transport of all sizes
		Local alteration of surface hydrology crossed by roads, resulting in increased sediment and increased soil erosion
		Changes in the level of the water table
		Sea Water Intrusion and flood control infrastructures will cause further ecological and habitat dis-connectivity in Mekong delta

(continued)

Table 2.10 (continued)

Aspects	Indicators/Variables	Foreseeable 20 years 2030 with LMB dams
	From UMB	75–80 % reduction (90–20 Mt/year) Half the amount of nutrients
		Medium size sediment to Kratie-PP –TLS decrease to 0 by 2050
		Fine sized sediments reduced by another half at all locations
		Nutrient reduced by another 50 %
	Within reservoir area	Formation of deltaic type deposits at the head of each of the reservoirs
		Middle and lower parts of each reservoir will sediment during flood events, associated with reduced velocities/gradients
		Sedimentation and deltaic formation at head of reservoirs
		Sedimentation in reservoirs
	Downstream and critical deep pool and habitat	Erosion starts from Chiang Sean and progressively downstream but slowdown by deep pool
		Increased down cutting and bed and bank erosion in Vientiane-Pakse reaches
		Riverbed and bank erosion Vientiane-Pakse reaches, and Pakse-Kratie. Changes in Thalweg border lines between Lao PDR and Thailand
		Bank and bed erosion starts to be felt at Kratie to PP reaches
		No more supply of sand sized sediment to Mekong delta (not enough stream power for suspension)
		335 deep-pools continue to function, but their long-term functioning may affect
		48 % of them are lost. (About 70 % on Chiang Saen- Vientiane reaches)
	Mekong delta	Reduce in floodplain sedimentation
		§165 million tonnes per year down to 36 % in 20 years
		Stability of the delta shaping processes due to loss in sediment deposition, which are potentially further exacerbated by sea level rise
		Greater instability and erosion of channel
		Some sections of the dykes would have to be even relocated to further land due to the seashore erosion
		Rises in sea water levels combining with increased water consumption within the Mekong delta and upstream will push saline water further inland
		Higher embankments and strengthening of dikes, roads and sluice gates for salinity intrusion will lead to hydrological and morphological changes and variation (obstructions to flood drainage, increasing flood peaks, and increase sedimentation and erosion) on both sides of the Vietnamese and Cambodian border

Estimated increases in mean monthly flows due to proposed dam releases are anticipated for at least the 3 months of February to April. But in early dry season (December) most stations downstream from Vientiane are expected to experience reduced flows due to the combination of dam operations and the high demand for irrigation. If the planned large scale diversions in Thailand materialize, further reduction in base flow is expected. The dry season flow change would be more marked with the projected increase of the dry season irrigation in the area further downstream of Kratie, but may be compensated slightly by hydropower reservoir operations in the Vietnam Central Highlands.

Areas immediately below (40–50 km) reservoirs can experience up to 3–6 m of daily flow fluctuations from peak operations, and emergency releases to avoid dam failure due to flooding can cause flash flooding in the immediate downstream stretches.

Moreover, without river specific and calibrated water balance accounting the projected increase in dry season flow from upper mainstream stretches may not directly benefit any planned diversions from the tributaries. For instance, intra-basin diversions will cause significant reductions in dry season flows in source tributaries such as the Nam Ngum and Nam Lik. Opportunities exist for both positive and negative transboundary impacts, at this time mostly constrained to Lao PDR and Thailand. Positive impacts are related to the benefits of access to additional irrigation water, assuming effective infrastructure and irrigation efficiencies are in place. However wet season reservoir operation poses the risk of flood hazard and additional irrigation water can mobilise soil or groundwater salts. Mainstream reaches by-passed by the diversion are exposed to additional, cumulative risks, especially in drier years. A constant diversion potentially causes water shortages during the critical dry season months of February to May, reducing tributary contributions that maintain minimum environmental flow in the Mekong mainstream.

5.3.3 Expected Changes in Flooding Conditions

The onset of dry to flood season transition will be significantly reduced by 7–8 weeks earlier in Chiang. Saen, 2–4 weeks earlier upstream of Pakse and 1 week at Kratie. This change will see a reduction in the important freshwater '*spates*' which drive many ecosystem functions such as fish spawning, lave dripping and fish migration.

Substantially reduced irrigation needs and limited water storage capacity restrict the current level of wet season water diversions. The impact of dam and irrigation developments on flood conditions will vary at different key points of the Mekong River. Proposed mainstream dams in the LMB will provide limited flood protection to the location immediately downstream (MRC 2010b, c, d). Records from around the world indicate that revenue neutral flood protection tends to be neglected in multipurpose projects (Regional POE 2010).[11]

Reduced flood season peak flows are predicted to reduce the extent and duration of floodplain inundation and contribute to bank erosion on the critical stretches and infilling of deep pools (MRC 2009). The average annual reduction in flooded area is estimated at 7 % (309,000 ha), expected to be smaller in wet years and larger in dry years. The greatest area of reduction occurs in Cambodia (142,000 ha), Lao and Thailand (17 % and 19 % reduction respectively) (MRC 2010d; MRC 2009).

5.3.4 Reverse Flow in Tonle Sap and Impact on Water Level and Inundated Area, Duration and Timing

The changes in the Mekong flood pulse are expected to have transboundary impacts, particularly for the ecology of the Tonle Sap Lake and Mekong delta floodplains (MRC 2010c). The inundation of the Tonle Sap Lake will reduce by 5–10 % (500–600 km^2). However, seawater intrusion prevention measures in the Vietnamese delta will contribute to increased and longer duration flooding in both Cambodia and the delta. Extended flood drainage duration for the entire delta is anticipated, leading to flow stagnation and attendant reductions in water quality.

The Tonle Sap reverse flow will commence at least a week sooner, and the average days will be about 8 days shorter. Dry season inundated area is estimated to increase by 5–8 % transforming the affected seasonal terrestrial ecosystem into permanent aquatic. These changes will reduce the flooded area, flood depth and sediment inflow into the lake, influencing ecosystem and farming productivity, fish migration and sediment flushing capacity.

5.3.5 Change in Water Quality (Turbidity and Relevant Quality Parameters)

Dam sediment trapping will reduce Mekong sediment transport and attendant turbidity by 75 %. The predicted expansion and consolidation of irrigated agriculture

[11] Most of the dams have not been designed to take into consideration major natural disasters such as earthquakes or floods. Lessons from the typhoon Ketsana (end of September 2009) show that mismanagement can cause misery to vulnerable communities. Even where they have been considered, the imperatives of maximizing revenue obliged the dam operators to keep a reservoir as full as possible (and thus maximize hydropower generation as well as irrigation water availability). In recent years a number of earthquakes have taken place across Asia resulting in impacts to numerous dams (Mongabay.com 2008; Brewer 2008; Vijay and Ramesh 2005; Hough and Martin 2001). For example, the 12 May 2008 earthquake in Sichuan province of China (7.9 magnitude) seriously damaged hydroelectric dams and caused major social and economic losses. Sixty-nine dams were in danger of collapse, 310 were at "high risk," and 1,424 posed a "moderate risk" (Brewer 2008). China said it would spend more than $1.3 billion per year fixing vulnerable dams, many of which were poorly constructed (Mongabay.com, 2008; Brewer, 2008). USGS Earth Quake Hazards Project reported two major quake measuring 4.7 on the Richter scale in late February 2011 and 6.1-magnitude quake in 2007at the proposed mainstream dams near Xayaboury, Lao PDR.

will also lead to an increase in use of chemical fertilizer to offset losses in sediments, increasing variable production costs.

As a corollary a considerable basin-wide increase in herbicide and pesticide/fungicide use (75 % and 59 %) is expected and an overall increase in nutrient loading from irrigated areas (85 % (N) and 100 % (P)). The increases will be largest in the Mun – Chi Basin, the 3Ss Basin (agricultural run-off and mining operations) and the Tonle Sap Basin Increased urbanisation along the mainstream and key tributaries will lead to an 33 % increase in N and P nutrient loading from wastewater discharges.

A substantial reduction in water quality is expected as a result of large-scale mining in the Bolaven plateau. Vietnamese government investments in higher embankments, sluice gates and strengthening of dikes and roads will accentuate water quality reductions by increasing the level of flow stagnation and reduced flushing capacity.

5.3.6 Change in Salinity Intrusion: Extent, Duration, and Concentration

Vietnamese irrigators in the Mekong delta are demanding increasing volumes of water during the dry season period of April and May (MRC 2002) Predicted sea level rises and subsequent salinity intrusion will further exacerbate water shortages in periods of extreme drought. The interaction of predicted dry season flows from dam releases and salinity intrusion remains uncertain but potentially a significant determinant of aggregate agricultural production in the delta.

The return flow from the Tonle Sap during the early stage of dry season (December to February) substantially contributes to the downstream flows at Chaktomuk and eventually the Mekong delta. confirming the important role of the Tonle Sap Great Lake as a regulator of Mekong flows. A drastic change to the flow level at the Mekong delta will be expected if the flow into and from Tonle Sap is drastically reduced or delayed by upstream dams operations and diversion. Figure 2.9 illustrates the relative delta flow contributions of the Tonle Sap and upper Mekong.

The delta irrigation volumes registered in 2000 constituted 76–81 % of the available flow in the critical dry season months of April and May, correlated with the increased incidence of seawater intrusion (World Bank 2004). With a projected growth in irrigation water use in April and May serious water shortages and increased competition among water uses is anticipated, as current irrigation demand is already approaching flow availability.

Earlier MRC commissioned studies estimated that during the low flow period the Mekong delta flow requires at least 1,500 m^3/s to prevent severe sea water intrusion (Daming 1997; SMEC 1998). The MRC and Vietnamese Sub-Institute for Water Resources Planning estimated the water requirements for crops in the Vietnam's Mekong delta (year 2000) was at around 1528–1018.7 m^3 per second during February until May, when the observed mean discharge at Phnom Penh, Cambodia was measured between 1,984 and 2,769 m^3/s during that same period (MRC 2002).

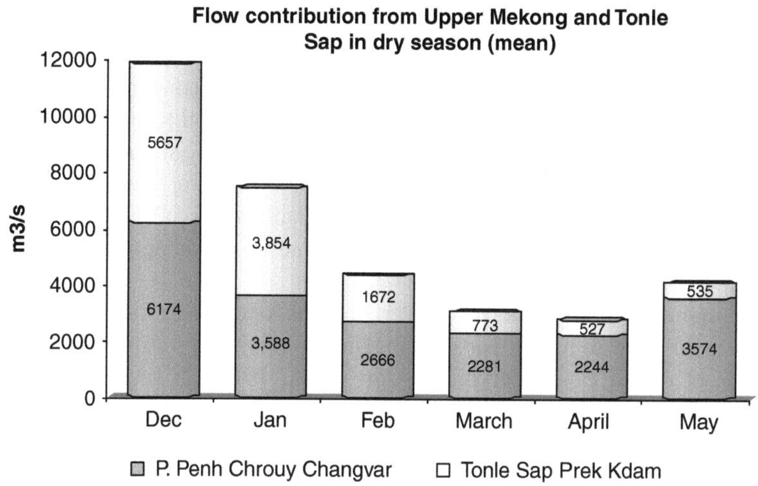

Fig. 2.9 Chaktomuk average monthly minimum flow from upper Mekong and Tonle Sap Rivers

In order to satisfy the water requirements of predicted irrigation expansion and to prevent sea water intrusion in the delta area, additional dry season flows will be required in March, April and May either from locations above Kratie or from the Tonle Sap Great Lake.

5.3.7 Geomorphologic Changes

Despite some proponent claims of benign run of river impoundments, large dams in particular are predicted to reduce sediment transport and associated nutrients, and disconnect wetlands from the river system.

Bravard and Goichot 2010 refer to four dimensions of geomorphologic change in a river system: the longitudinal, lateral, vertical and time. The downstream impacts on the longitudinal profile will trigger localized riverbank erosion and indirect effect of a higher level process that will influence the lateral dimension of the river geomorphology. Erosion of the riverbed, a reconstitution of bed load, and a reduced river bed slope are predicted due to dissipation of river energy. Consequences for the populations living on the river banks (immediately downstream in the area around Vientiane, and to a greater extend on the stretch downstream from Kratie to the Vietnamese border) will be much more critical than anticipated in the current scenario assessment (Bravard and Goichot 2010).

Incision of the river bed in alluvial stretches from Vientiane to Paske, and downstream from Kratie from a sediment hungry river are also another potential impact and will result in water table and riparian vegetation changes. Dams directly upstream of these areas will reduce bed loads and rapidly lead to significant loss of habitat diversity and reduced fisheries productions.

5.3.8 Barrier Effects and Disconnectivity

The construction of the 11 LMB mainstream projects will convert 55–66 % of the 1,760 km river distance between Chiang Saen (Thailand) and Kratie (Cambodia) to a series of managed reservoirs. From 9,000 to 150,000 ha of garden and agricultural land will be inundated in dam reservoir sites (54 %) in Chiang Saen to Pakse, and a 34 % reduction of Lao PDR wetland 18 % in Thailand, and 2.4 % in Cambodia. This change is expected to cause substantial changes in key ecological habitats and water dependent productivity.

With mainstream hydropower power projects operating, there will reduced velocity to suspend sediment particles leading to enhanced sedimentation and the formation of deltaic type reservoir deposits (MRC 2010d; ICEM 2010). Reliable estimates of reservoir sedimentation rates have been hampered by limited sediment size gradients data, except for fine sized particles, measured in the Mekong mainstream since the 1960s.

The main impact of built infrastructures in the floodplains and across Mekong basin rivers will be the alteration of surface hydrology, modified distribution of sediment loads, increased soil erosion and modified flood patterns.

5.3.9 Critical Deep Pool and Habitat

The development will also potentially reduce upper Mekong Basin fined sized sediment transport from 90 to 20 Mt/year at Chiang Sean, and from 165 to 88 Mt/year at Kratie. The reduction will result in a significant loss of floodplain nutrients in floodplains, making deep pool fish habitat at most stretches less productive, reducing fish capture harvests.

6 Conclusion

The water issues in the Mekong Region are closely related to the unequal spatial and temporal distribution of flow, and the lack of robust institutional coordination and well-informed decision making for water resources development. Much of the runoff occurs in flood events or remains inaccessible because of remote locations. The proportion of runoff and river recharge required to maintain non-extractive instream functions and other social and ecological services remains poorly defined.

Despite many Mekong Region countries being endowed with relatively high amounts of renewable water resources per capita, a number of locations currently face a series of critical water issues such as: water shortages and growing water demand; intensive salinity intrusion in the Mekong delta; fisheries reductions and degradation of natural habitats; increased frequency and intensity of floods and drought; and intensification of sectoral competition within and among the Mekong countries.

Water and ecosystem productivity and integrity in the Mekong River Basin are closely linked to the seasonal flow pattern of the river, typified by a wet season flow up to 10 m higher than the dry season. Policy deliberations will need to assess potential water sector changes based on those natural fluctuations, focusing on hydrological changes, and geomorphologic changes that modify ecosystem sustainability and sustainable livelihood. The multi-sector demand and associated dam and diversion interventions remains the key trigger of changes in the Mekong Region for the next 30 years, associated with significant transboundary implications. The extent of hydropower development remains as the primary locus of political attention and negotiation at national and supra national scales. The impact of large scale mining and processing, road and railway infrastructures, and salinity intrusion prevention and flood control structures in the floodplains and Mekong delta will have significant transboundary implication, albeit generally manifest at the local scale. The magnitude of impacts will be dependent on the number and location of dams that will be actually built and how dams are designed, built and operated.

The positive and negative impacts from selected hydropower plans and projects are well documented, although interpretation remains highly contested. However, uncertainties remain relating to the nature, extent and distribution of benefits and impacts on vulnerable communities and water dependent interest. The characteristics of the development trajectory are likely to have a significant impact on the nutritional status of the poor given the extent of the expected reduction in fisheries. This may be aggravated by increases in urban poverty from increased rural–urban migration due to a declining natural resource base and may have knock-on implications for urban poverty rates and an increase in food prices. Those displaced people losing agricultural land would not be the same people who would benefit from improved irrigation opportunities. Small scale land holders may not benefit from medium and large irrigation schemes (ICEM 2010).

It is challenging to provide defensible economic values to the loss to paddy production, fishery and other wetland services. As a general rule, hydropower developments externalise costs, transferring a large proportion to affected communities and undervalue the modification to ecosystem goods and services. Significant mitigation and compensation costs due to these developments, which are expected to be both financially and politically extensive, need to be included in the economic valuation.

There are a number of limitations and caveats regarding the analysis discussed in the Chapter. It is not possible to assess the impacts of the numerous dam and irrigation projects due to limited availability and informally collated data. It is important to note that the current hydrological changes modelled by the MRC relied on to derive these conclusions are built on a number of rough assumptions. Furthermore, in spite of some attempts to create a framework for assessing the impact of various development projects in the MRB, there has been no "cumulative assessment" carried out in the Mekong River Basin. The absence of such a comprehensive impact assessment and an underpinning monitoring system in the Mekong Region highlights the concern for severe cumulative impacts on resource sustainability and livelihoods. However, there is general agreement that these major development projects will substantially modify the hydrological regime and associated water elements in both an incremental and a cumulative manner.

References

ADB, and SEI. 2002. Strategic environmental framework: Integrating development and environment in the transport and water resources sectors, Vols. II, II and IV. Manila: Asian Development Bank and Stockholm Environment Institute.

ADB (2009), 'Annual Report of the Community of Practice on Water', (Manila: Asian Development Bank).

ADB – RETA 40082. 2011. 3S Technical sheets of key topics no 8 large scale infrastructure development in the 3Ss 8b – mining development. Accessed on line on 16 Feb 2011 at http://reta.3sbasin.org/index.php?option=com_content&view=category&layout=blog&id=142&Itemid=178&lang=en.

Baran, E. 2005. *Cambodian inland fisheries – facts, figures and context*. Penang: World Fish Centre.

Baran, E., P. Starr, and Y. Kura. 2007. Influence of built structures on Tonle Sap fisheries Cambodia National Mekong Committee and the WorldFish Centre. Phnom Penh. 44.

Bravard, J. P., and M. Goichot. 2010. Ensuring ecosystem integrity and ecosystem services: A critical review of the BDP (Basin Development Plan) environmental assessment. Critical review by International NGOs at the BDP2 3rd Stakeholders Forum, Vientiane 29–30 July 2010.

Catch and Culture. 2005. Fisheries research and development in the Mekong Region, Vol. 11, no.1, issue of May 2005, fishery programme. Vientiane: Mekong River Commission Secretariat.

CSIRO. 2011. Measuring forestry's impact on water availability. Accessed on line on 18 Feb 2011 at http://www.csiro.au/news/Measuring-forestry-impact-on-water-availability.html.

Daming, H. 1997. Sustainable development of Lancang-Mekong River Basin and integrated multi-objective utilization research of water Resources, *Journal of Chinese Geography* 7(4):9–21.

Dore, J., and X. Yu. 2004. Yunnan hydropower expansion: Update on China's energy industry reforms and the Nu, Lancang and Jinsha hydropower dams. Chiang Mai University's unit for social and environmental research, and Green Watershed.

Dosch, J., and O. Hensengerth. 2005. Sub-regional cooperation in Southeast Asia: The Mekong Basin, Brill academic publishers. *European Journal of East Asian Studies* 4(2): 263–286.

FAO. 2005. Increasing the contribution of small-scale fisheries to poverty alleviation and food security, *FAO technical guidelines for responsible fisheries*. No. 10. Rome: FAO.

Fox, J.M., J.B. Vogler, O.L. Sen, A.L. Ziegler, and T.W. Giambelluca. 2012. Simulating land-cover change in Montane Mainland Southeast Asia. *Environmental Management* 49(5): 968–979.

Grupo Guayubira. 2011. Testimonies about the impacts of large-scale tree plantations on flora and fauna, Montevideo – Uruguay Accessed on line on 18 Feb 2011 at http://www.guayubira.org.uy/english/florafauna.html.

Halcrow Group. 2003. Water utilization project component A: Hydro-ecological report (discussion paper 11). Phnom Penh: Mekong River Commission Secretariat.

Halcrow Group. 2004. Development of basin modelling package and knowledge base (WUP-A), DSF 650 technical reference report appendix A 1–6. Phnom Penh: Halcrow Group Limited, for Mekong River Commission Secretariat.

ICEM. 2010. MRC strategic environmental assessment for hydropower on the Mekong mainstream impacts assessment (opportunities and risks), discussion draft, MRC Initiative on Sustainable Hydropower (ISH). Vientiane: Mekong River Commission Secretariat.

Jiabao, W. 2005. *Opening statement – A stronger partnership for common prosperity*. Kunming: 2nd Greater Mekong Sub-region (GMS) Summit.

Keskinen, M., and M. Kummu. 2010. *Impact assessment in the Mekong – Review of Strategic Environmental Assessment (SEA) & Cumulative Impact Assessment (CIA), improving Mekong water allocation project (PN67) Challenge Program on Water and Food (CPWF)*. Finland: Water and Development Research Group, Aalto University.

Lazarus, K. 2009. *In search of aluminum: China's role in the Mekong Region*. Cambodia: Heinrich Böll Stiftung Cambodia, World Wild Fund and International Institute for Sustainable Development.

Menne, W. 2004. Timber plantations in Swaziland: An investigation into the environmental and social impacts of large-scale timber plantations in Swaziland, Timberwatch coalition member, commissioned by the World Rainforest Movement (WRM).

Miyazawa, N., K. Sunada, and S. Pech. 2008. Bank erosion in the Mekong River Basin: Is bank erosion in my town caused by activities from my neighbours? In *The modern Mekong myth book*, ed. M. Kummu, M. Keskinen, and O. Varis, 19–26. Helsinki: Finnish Academy of Sciences, Helsinki University of Technology (TKK).

Miyazawa, N., S. Kengo, and S. Pech. 2008. Book chapter: Bank erosion in the Mekong River Basin: Is bank erosion in my town caused by activities from my neighbours? In: *The modern Mekong myth book*, ed. The Matti Kummu Marko Keskinen Olli Varis, 19–26. Finnish Academy of Sciences, Helsinki University of Technology (TKK).

Molle, F., and P. Floch. 2008. The "Desert bloom" syndrome: Irrigation development, politics, and ideology in the Northeast of Thailand. Working paper. Mekong program on water, environment and resilience, Institut de Recherche pour le Développement, International Water Management Institute. M-POWER, Chiang Mai. http://www.mpowernet.org/mweb.php?pg=92

Molle, F., T. Foran, and M. Kakonen. 2009. *Contested waterscapes in the Mekong Region: Hydropower, livelihoods, and governance*. London: Earthscan.

MRC. 2002. Basin development planning regional sector overview – agriculture and irrigation report. Phnom Penh: MRC Secretariat, Basin Development Plan Programme.

MRC. 2003. State of the basin report. Phnom Penh: Mekong River Commission Secretariat. 316 pages. ISSN 1728:3248.

MRC. 2005. The MRC basin development plan – scenario for strategic planning, BDP library. Vol. 4, Vientiane, Lao PDR: Mekong River Commission Secretariat.

MRC. 2009. The modelling the flow of the Mekong, MRC management information booklet series no 3. Vientiane, Lao PDR: Mekong River Commission Secretariat.

MRC. 2010a. State of the basin report. Vientiane, Lao PDR: Mekong River Commission Secretariat.

MRC. 2010b. Assessment of basin-wide development scenarios, basin development plan programme, Phase 2. Vientiane, Lao PDR: Mekong River Commission Secretariat.

MRC. 2010c. IWRM-Based basin development strategy for the lower Mekong basin – first complete draft – 15 Sept 2010. Vientiane, Lao PDR: Mekong River Commission Secretariat.

MRC. 2010d. Impacts of climate change and development on Mekong flow regime. First assessment – 2009. MRC Technical paper no. 29. Vientiane, Lao PDR: Mekong River Commission Secretariat.

MRC. 2010e. Technical note 2 – Hydrological assessment, Vientiane, Lao PDR: Mekong River Commission Secretariat.

Nikula, J. 2008. Is harm and destruction all that floods bring? In *The modern Mekong myth book*, ed. M. Kummu, M. Keskinen, and O. Varis, 19–26. Helsinki: Finnish Academy of Sciences, Helsinki University of Technology (TKK).

Papademetriou, M.K. 2000. *Rice production in the Asia–Pacific region: issues and perspectives*, Food and agriculture organization, regional office for Asia and the Pacific, www.fao.org.

Pech, S. 2010. Cambodian and Mekong water resources governance. In *Transboundary resources and environment in mainland Southeast Asia*, ed. Sato J. Institute for advanced studies on Asia, The University of Tokyo, Japan.

Pech, S., and K. Sunada. 2006, China natural resources demand: Opportunities and challenges for Mekong sub-region, Processing of regional workshop "China in mainland Southeast Asia: Flying with the Dragon", Chiang Mai 17–19 Oct 2006.

Pech, S., and K. Sunada. 2008. Population growth and natural resources pressures in Mekong River Basin. Ambio – *A Journal of the Human Environment* XXXVII(3): 219–224

Raskin, P.D., and E. Kemp-Benedict. 2002. Global environmental outlook scenario framework. *Background paper for UNEP's third global environmental outlook report.* Boston: Stockholm Environment Institute.

Regional POE. 2010. Review of the Mekong river commission's basin development plan programme Phase 2. Vientiane: Panel of Experts Report prepared for the Mekong River Commission Secretariat, M-POWER.

Revenga, C., and G. Mock. 2000. Pilot analysis of global ecosystems: Freshwater systems and world resources 1998–99. http://earthtrends.wri.org/features/ view_feature.

Sanyu. 2004. Nam Ngum water management project for Vientiane plain of Lao PDR and northeast Thai region, by Sanya Consultants Inc. SCI, Japan.

Sarkkula et al. (2010). Juha Sarkkula, Jorma Koponen, Hannu Lauri, Markku Virtanen. Origin, fate and impacts of the Mekong sediments. Mekong River Commission, Vientiane, Lao PDR.

Shigeko. 2009. *The Mekong river delta.* Kokon Shoin Publishers

Smajgl, A., J. Ward, T. Foran, J. Dore, and S. Larson In review. Visions, beliefs and transformation:Methods for understanding cross-scale and trans-boundary dynamics in the wider Mekong Region. Global Environmental Change.

Smakhtin, V., C. Revenga, P. Döll, and R. Tharme. 2003. Giving nature its share: Reserving water for ecosystems. In Putting the water requirements of freshwater ecosystems into the global picture of water resource assessment, ed. A. Moy. Accessed on 6 Feb 2011 at http://earthtrends.wri.org/features/view_feature.php?theme = 2&fid = 38.

SMEC. 1998. Final report water utilization programme preparation project. Bangkok: SMEC/MRC Secretariat.

Thompson, R. 2010. Regional and international country experiences lessons learned, http://www.un.org.kh/undp/international-conference-on-mining/international-conference-on-mining#keynote-presentations.

UN. 2007. World population prospects: The 2006 revision, population division of the department of economic and social affairs of the United Nations Secretariat.

UNEP/GIWA. 2006. Global international waters assessment: Mekong River – GIWA regional assessment 55, University of Kalmar for United Nations environment programme.

Van Zalinge, N.P., 2002. Update on the status of the Cambodian inland capture fisheries sector with special reference to the Tonle Sap Great Lake, Mekong Fish Catch and Culture, Vol. 8 no. 2.

WRI. 2011. Earthtrends environmental information. Washington, D.C.: World Resource Insititue. http://www.wri.org/project/earthtrends/.

World Bank. 2004. Mekong Regional water resources assistance strategy – modelled observations on development scenarios in the lower Mekong basin. Vientiane: World Bank.

World Bank and ADB. 2006. Future directions for water resources management in the Mekong River Basin. Joint working paper. Vientiane: World Bank and the Asian Development Bank, 65.

Xu, A., and D. Moller. 2003. Hydropower development plan set for Lancang River. In http://www.china.org.cn/english/2003/Dec/82505.htm. Accessed on 10 Feb 2011.

Ziegler, A.D., J.M. Fox, and J. Xu. 2009. The rubber juggernaut. *Science* 324: 1024–1025.

Chapter 3
Food Security in the Wider Mekong Region

David Fullbrook

1 Methodological Background

1.1 Introduction

The wider Mekong Region[1] can appear to be awash with food, as a walk through the colour and pungency of a market in almost any reasonably prosperous town will show. That about half the rice traded on world markets was cropped in the paddies of these lands, especially the Chao Phraya Plain and Mekong delta, might also suggest food is abundant and secure. For most people it is, but for a considerable minority it is not. Furthermore, the security of the resources employed by farmers and fishers to feed their communities and people around the world is in question because of the world's growing appetite for energy and materials.

Investors, with the permission of governments, are putting labour and capital to work, fundamentally rearranging the natural energy and ecological infrastructure upon which food production depends. Dams are blocking rivers, rubber and acacia are marching across the land. This may not be a bad thing. Sometimes investment can under the right conditions produce a net benefit for well-being, in which case society develops. However, without careful planning and management it is by no means guaranteed that well-being, which at the very least requires all have sufficient food for good health, will improve.

The question is whether investment and industrialisation of the wider Mekong Region is positive or negative for maintaining food production to provide food security for its citizens, and possibly others overseas through trade. It is not a simple question because of the scale and speed of change and the possibility of the consequences behaving non-linearly and interacting to amplify impacts or produce

[1] The Asian Development Bank defines the region as Cambodia, China (Guangxi and Yunnan), Lao PDR, Myanmar, Thailand and Vietnam embracing the major rivers Chao Phraya, Irrawaddy, Mekong, Salween and Red.

A. Smajgl and J. Ward, *The Water-Food-Energy Nexus in the Mekong Region: Assessing Development Strategies Considering Cross-Sectoral and Transboundary Impacts*,
DOI 10.1007/978-1-4614-6120-3_3, © Springer Science+Business Media New York 2013

outcomes *ignoramus et ignorabimus*. A robust answer will come in the future from either a comprehensive research project or the turn of events. In the meantime the question is given a modest, and inevitably, imperfect treatment in this Chapter through a cumulative assessment of the impacts of large dams on the mainstream of the Mekong River, diversion of water from the Mekong River, rising seas, industrial crops, transportation and mining. At best this is a starting point for further reflection on the challenges and possibilities of nurturing sustainable and resilient agriculture to ensure universal food security in the wider Mekong Region while also unlocking opportunities for economic development to improve wider well-being.

1.2 Food Security

Food security is a state of experience for individuals and communities. It is described and quantitatively measured to assess suffering and plan remedies. How a state of food security comes about is a complex process. An individual's state of food security will partly depend upon their knowledge and preferences. However, fundamentally the change and coordination of the use of natural resources plus the distribution of wealth through social institutions particularly the economy determines whether the context is positive or negative for strong food security.

The World Summit on Food Security in 2009 declared "Food security exists when all people, at all times, have physical, social and economic access to sufficient, safe and nutritious food to meet their dietary needs and food preferences for an active and healthy lifestyle" (WSFS 2009, p. 1). The definition, like many others in circulation, describes a state in which people have sufficient quantity, or energy, and quality, or vitamins and nutrients, of food to live in good physical and spiritual health today and tomorrow. People suffer weak food security[2] when they are exposed acutely or chronically to insufficient energy or nutrients, or both. Exposure to insufficient energy or nutrients is not necessarily synchronous illustrating the complexity of food security (Fig. 3.1).

The state of food security has four dimensions (FAO 2006):

- Availability – the quantity and quality of food supply is sufficient for human needs
- Accessibility – people exercise control over adequate means, such as land, inputs, money, etc., to acquire food they require for good health
- Utilization – the context in which food is prepared, consumed and disposed is supported by clean water, dietary knowledge and safe sanitation
- Stability – food supply is resilient through acute shocks and chronic or cyclical events ensuring sustained availability and access at all times (Fig. 3.2).

[2]Weak food security is often referred to as food insecurity, however for clarity and consistency it is preferable to think of food security operating on continua from weak to strong

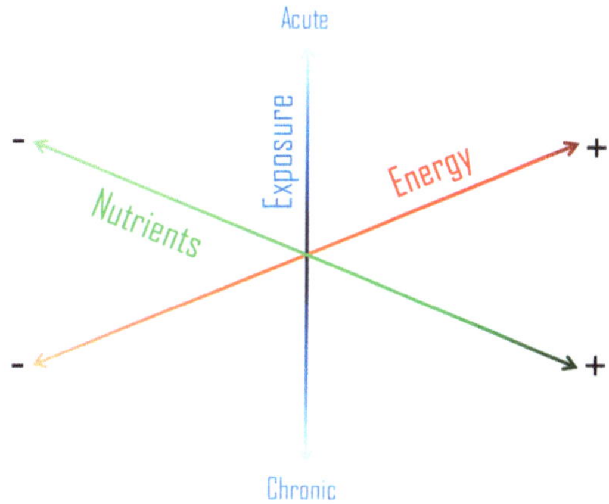

Fig. 3.1 Conceptual experience of food security

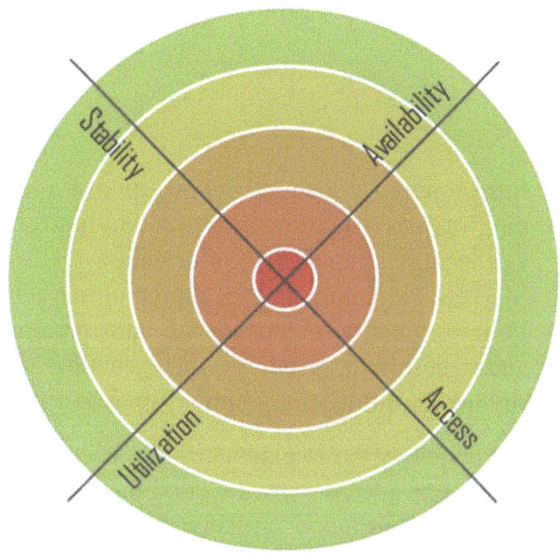

Fig. 3.2 Dimensions of food security

Food security is the product of the unique context of a community, country or region. The quantity and quality of natural resources made available for producing food relative to the requirements of the population is an important relationship which is the focus of this analysis which introduces the concept of enabling factors. Finding equilibrium between resources and requirements for an optimal outcome in terms of strong food security is the product of the interactions between enabling factors which may vary in scope and intensity with each situation at a particular time. Key enabling factors might include:

- Demography – size, growth, age, skills, urban–rural ratio, wealth distribution, etc.
- Resources – quantity and quality of natural resource endowments for agriculture as well as impacts of using resources which compete for fundamental resources such as land and water

Fig. 3.3 Enabling factors of food security

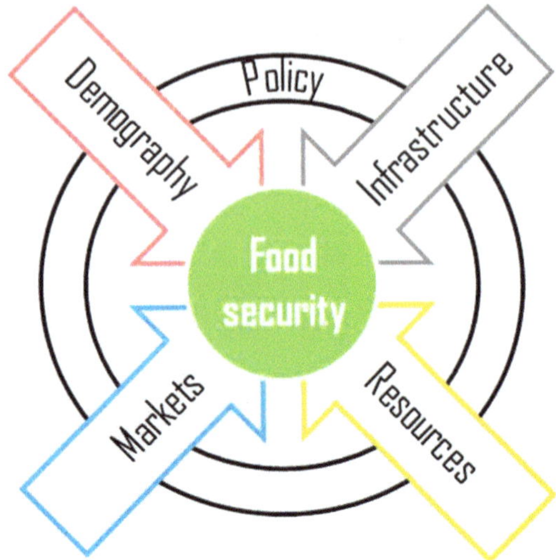

- Infrastructure – quantity and quality of ecological infrastructure, hard infrastructure, such as irrigation, transport and food storage, plus soft infrastructure, such as education, knowledge, information
- Markets – structure and regulation of markets for food, resources, labour, etc.
- Policy – quality, independence and impartiality for coordinating employment and trade-offs in resources, infrastructure and markets relative to the food security needs of the demography (Fig. 3.3).

Public policy has to play a key role because although the consumption of food is generally considered a private good the security of food is very much in the public interest, a public good unlikely to be guaranteed by market forces alone in the absence of collective action by society. The challenge is establishing policy to secure food without creating unsustainable costs, liabilities or interests within the food system, which run counter to the goal of strong food security.

The exercise of preferences, priorities and power influences the extent to which society experiences strong food security. Until the food shocks of 2008 few countries, China being a notable exception, appeared to take food as seriously as other elements of national security such as military power or energy (Fullbrook 2010). A state of strong food security is vulnerable when other goods demanded in an economy and facilitated by public policy take precedence, whether intentionally or unintentionally, over the provision of quality food in sufficient quantities at affordable prices.

As food is the most fundamental human need it may be assumed, at least in principle, that the economy's primary goal is to satisfy the need for food before all else. Therefore, when food security is weak or inadequate the economy has failed or is underdeveloped. That inadequate food security persists despite sufficient food resources in the wider Mekong Region and the world while other needs and wants are being met for some suggests asymmetric power distributions which enable some groups to exert control over food and other resources at the expense of others, who through their hunger and malnutrition in effect subsidise the richer food and other goods enjoyed by those with more power. Poor food security indicates policy may be imbalanced, warped by special interests, or public administration suffers from insufficient knowledge, skills and authority. Ensuring food security raises issues equity, justice and fairness which are taken up by the United Nations Special Rapporteur on the Right to Food.[3]

Securing food raises questions over whether the cheapest food is necessarily the most effective way to assure availability of food at all times. One of the dilemmas of food security is between narrow economic efficiency, which would favour letting comparative advantage work out security through markets and trade, or potentially higher costs, but probably greater security, of supporting domestic production. Since neo-liberal policy came into vogue in many countries and international institutions the former has come to the fore. However, the food price shocks of 2008 and stubborn persistence of higher prices along with perplexing challenges and uncertainty, such as climate change, have been cause to reconsider domestic production. Of course a few countries, notably China, have long considered maintaining domestic food production critical to national security.

An assessment of the context of food security will necessarily have to take a broad view of the elements of a situation and how they are or might interact. A long view is also almost certainly necessary because changes may play out over years and coordination of resources may take even longer to put in place the soft and hard infrastructure conducive to long-term strong food security.

Amid this complexity the concern here is with one aspect, how economic development affects the production and supply of food, especially when impacts flow across borders of the wider Mekong Region through the environment and economy.

1.3 Geography

The wider Mekong Region encompasses five major rivers under mostly tropical climates. Productivity of staples such as rice and protein from fish in the rivers and coastal waters is high. Thailand and Vietnam are among the world's leading exporters of food. Thailand is the leading supplier of rice for international trade, closely followed by Vietnam. Indeed almost a quarter of the rice on the world market is grown in the paddies of the Mekong delta (Fullbrook 2012).

Nevertheless despite the cornucopia, food security is neither currently ubiquitous nor assured. Aside from policy, markets and poverty, natural conditions, such as

[3]http://www.srfood.org/

water scarcity on the Khorat Plateau in northeastern Thailand or the dry zone of central Myanmar, events such as floods and climatic cycles, associated for example with the El Niño Southern Oscillation climate pattern, can affect the production and supply of food locally and at times at large scale.

Economic development also exerts considerable pressure on the land with consequences for food production. For example, food production is displaced by production of industrial materials or lost to urbanization and factories. Cities or industry can place additional demands on aquifers reducing groundwater supplies for farming.

1.4 Development

The wider Mekong Region is experiencing great change as industrialization proceeds apace accompanied by improvements in transportation and communications, which expand markets to embrace ever more communities more comprehensively erasing subsistence. Lao PDR and Myanmar rank among the least developed countries in the world, with indicators typical of many countries in Africa. By contrast, China and Thailand are comfortably in the realms of middle income states. China, Thailand and Vietnam are quickly becoming urbanized societies. Cambodia, Lao PDR and Myanmar remain largely rural with considerable food security challenges which have drawn long-term assistance from the World Food Programme.

Development brings great shifts in the wider Mekong Region. People are migrating, some permanently others seasonally, from farms to work in factories, shops and offices. They no longer depend on themselves for food but on markets, in other words the population of net food buyers is expanding. If prices go up or supplies are interrupted then they are vulnerable. Food may be more likely to flow to those who can pay most and not necessarily to those in greatest need. The food production and supply in the Mekong may not be evolving fast enough to keep up with the requirements of society migrating from agriculture to industry and from the countryside to the cities. Food security of urban society will increasingly depend on the development of a robust system of food production and supply which is resilient to acute shocks, such as extreme weather, and chronic change, such as climatic variation.

Worldwide, prosperity and urbanization are associated with higher demand for meats, such as pork and beef, which require much more land, feed and water to produce then poultry or fish. Where those resources come from and whether their use can be sustained poses trade-offs for food security. Furthermore, the less efficient, more intensive and more expensive food demanded by wealthier people can leave less grains or pulses for poorer people unless production is able to sustainably intensify or expand into marginal lands or displace forests and their ecosystem services which, among other things, support agricultural production. Thus prosperity changes tastes to produce another force, along with those of industry and urbanization in general, for changes in land use. Changes in prosperity and demand play out across a variety of scales, reaching across borders and ecosystems interacting with other great changes, considered by this analysis.

The diversity of development in the region leaves some countries in a position to exert greater demands on natural resources and shape the distribution of costs across borders and ecosystems with consequences of the ecological infrastructure, which supports food production and the exposure of people to the impacts. Large hydroelectric dams on the Mekong mainstream, primarily to power Thailand, are a case in point. The diversity and character of development of the region's states may conspire to shift, concentrate and perpetuate chronic food insecurity among weak, marginal communities while concurrently sacrificing parts of the ecological infrastructure with long-term acute and chronic consequences for food production.

Managing food security, which among other things requires comprehensive environmental management to secure and maintain ecological infrastructure, is a complex challenge requiring highly competent governance and public administration. Causes play out over years and decades, while the consequences can intensify over days or weeks and linger for years. In the wider Mekong Region, however, governance and public administration vary considerably from the relative competence of China to the severely underdeveloped in Cambodia, Lao PDR and Myanmar. Not only does this present problems for domestic food security but also for regional food security where a high level of confidence, cooperation and coordination are required to sustain ecological infrastructure and avoid externalities being transmitted across borders.

1.5 Approach

Data which informs the opinions of this assessment is drawn from published literature and statistical databases of international organizations and bilateral development agencies.

The assessment focuses on a regional perspective of the context of food security, suggesting impacts of development on the four pillars of food security as well as the ecological infrastructure which supports food production. Local examples are considered as cases which may be repeated more generally in similar situations to assess effects on food security. Impacts and consequences are assessed towards the horizon of the year 2030.

The approach employs deductive and inductive reasoning. It is not presumed that all trends and effects can be identified from what is currently known about development and food security in the region. In other words, the perspective is informed by an assumption that the future of food security is non-ergodic, in the economic sense, not least because of the uncertainties in public policy, the global economy and climate change.

2 Status of Food Security in the Wider Mekong Region

2.1 Food and Population

The Global Hunger Index[4] indicates all countries[5] of the wider Mekong region have seen national food security strengthen over the last two decades (Table 3.1).

However, the status of food security is still particularly weak in the least developed countries of Cambodia, Lao PDR and Myanmar. The situation is better in China, Thailand and Vietnam, which are all middle income countries with relatively well developed government capacity and infrastructure. Further improvements in food security in Cambodia, Lao PDR and Myanmar might therefore be at least partly determined by the pace of progress in development of government capability and infrastructure.

Data collected by the Food and Agriculture Organization illustrates the divergence between relative improvements and gains for households and communities. Absolute numbers of people suffering from weak food security has not changed markedly. The number of people in Cambodia and Lao PDR undernourished in 2006–2008 was little changed from 1990 to 1992. Myanmar may face a similar situation as a least developed country (Table 3.2). One explanation for the situation in Cambodia and Lao PDR appears to be rapid population growth.

However, improvements in the ratio of rice production to consumption in Cambodia and Lao PDR suggest several interacting reasons for persistent weak food security including limited distribution, affordability issues and bland diet lacking in nutrients (Table 3.3).

Population by 2030 is projected to grow by approximately 25% in Cambodia and Lao PDR, where the effects of weak food security, such as malnutrition and child stunting, are already widely suffered. In Myanmar and Vietnam although the population is projected to increase approximately 15% in terms of actual people the increases are substantially greater. Myanmar's population projection warrants caution because it is similar to middle-income Vietnam yet the country's development is on par with Cambodia and Lao PDR, which like many poor countries experience population rising faster than more developed countries. Rising populations increase demand for land for food as well as urbanization and industrialization. Their needs coincide with the uncertainties for agricultural production of climate change as well as land loss due to rising seas (Table 3.4).

Urban population has increased throughout the Mekong, almost doubling in Cambodia, China and Lao PDR. Although a middle-income country, China is urbanizing at pace similar to the least developed countries of Cambodia and Lao

[4]The Global Hunger Index is derived from the proportion of the population undernourished, prevalence of underweight among children younger than five, and proportion of children dying before age five http://www.ifpri.org/book-8018/ourwork/researcharea/global-hunger-index

[5]Sources of country data do not in general produce sub-national data for Yunnan, however as this province is around the middle in terms of development in China it is assumed that national data approximates conditions in Yunnan.

Table 3.1 Global hunger index for the wider Mekong region

	Cambodia	*China*	*Lao PDR*	*Myanmar*	*Thailand*	*Vietnam*
2011	**19.9**	**5.5**	**20.2**	**16.3**	**8.1**	**11.2**
2001	26.3	6.8	23.6	22.5	9.5	15.5
1990	31.7	11.7	29.0	29.2	15.1	25.7

Key: Moderate | Serious | Alarming | Extremely alarming

Table 3.2 Undernourishment, stunting and population

	Undernourished people				Child stunting
	2006-08		1990-92		2008
Cambodia	**3.6m**	*25%*	3.8m	38%	40%
China	**n/s**	-	n/s	-	-
Lao PDR	**1.4m**	*31%*	1.3m	31%	48%
Myanmar	**n/a**		n/a		n/a
Thailand	**10.7m**	*16%*	15m	26%	16% (2006)
Vietnam	**9.6m**	*11%*	21m	31%	1%

Source FAOSTAT

Table 3.3 Consumption, production, spending

	Rice in daily energy supply		Rice production/ consumption ratio		Food in total spending
	2005-08	1990-92	2005-08	1990-92	2009
Cambodia	**64.5%**	79.9%	**169.6%**	81.4%	70.8%
China	**26.0%**	30.7%	**101.2%**	101.7%	-
Lao PDR	**64.2%**	72%	**147.9%**	107.2%	51.7% (2003)
Myanmar	**n/a**	n/a	**n/a**	n/a	n/a
Thailand	**37.9%**	47.1%	**264.7%**	186.5%	39% (2000)
Vietnam	**58.9%**	71.8%	**154.4%**	122.1%	50.1% (2008)

Source FAOSTAT

Table 3.4 Population and poverty (Source FAOSTAT)

	Population			Urban population		Poverty	
	2030	2010	1990	2005–08	1990–92	Rural	Urban
Cambodia	*17.4 m*	**14.1 m**	9.5 m	**20.9%**	12.9%	34.5% (2007)	11.8% (2007)
China	*1.4b*	**1.4b**	1.1b	**42.6%**	28.6%	2.5% (2005)	2.0% (1998)
Lao PDR	*7.8 m*	**6.2 m**	4.2 m	**29.7%**	15.8%	31.7% (2008)	17.4% (2008)
Myanmar	*54.3 m*	**48.0 m**	39.3 m	**n/a**	n/a	n/a	n/a
Thailand	*73.3 m*	**69.1 m**	57.1 m	**32.9%**	29.6%	10.4% (2009)	3% (2009)
Vietnam	*101.5 m*	**87.8 m**	67.1 m	**27.3%**	20.6%	18.7% (2008)	3.3% (2008)

PDR. This reflects the diversity of development in China, so while coastal provinces are already highly urbanized the poorer inland provinces, including Yunnan in the wider Mekong Region, are still urbanizing. Poverty, which often exposes people to weak food security, remains concentrated in the countryside. However, as urban population will probably expand faster than rural population in the wider Mekong region poverty may also become a more urban phenomenon presenting different challenges for food security.

Overall, data indicates national food security is steadily improving in the wider Mekong region. However, this assessment is subject to several strong caveats. One, the quality of data may vary considerably between countries. Two, accuracy of data especially for the countryside may be considerably less than the cities, moreover the shift from rural to urban societies may be underestimated because many urban residents are undocumented and therefore invisible to some statistical assessments. Three, although data for China is assumed to approximate conditions in Yunnan the peripheral location and proximity to poorer territories give reason to consider that conditions may lie between the national average and conditions found in poorer neighbouring states of the wider Mekong region. Four, despite national improvements food security remains weak and persistent for many households and communities, for example, in rural Lao PDR "levels of chronic malnutrition are still very high" with every other child affected by stunting (WFP 2007, pp. 98–99). Five, national improvements may reflect reductions in chronic weak food security but do not necessarily imply reductions in vulnerability to acute deteriorations in food security. Six, high cost of food relative to wages leaves many people vulnerable to rapid swings in food security status.

The trend of improving food security is not certain to persist into the future. Aside from population growth maintaining absolute numbers experiencing weak food security, economic development, which has supported improvements in food security, introduces challenges. Several examples highlight some of the overlooked or emerging issues for the wider Mekong region.

Apichart Jongskul, secretary-general of the Office of Agricultural Economics, identified a vulnerability in Thailand's food security arising from agriculture highly dependent on imported inputs such as seeds, fertilizers and fuel and the prospect of a sharp contraction in the labour force as aging farmers are not being replaced (Keeratipipatpong 2011). Rural people, helped by improvements in education in some places in the wider Mekong Region, can take up skilled or unskilled jobs with incomes higher and more stable than farming because of urbanization and industrialization. What is unclear is whether trends in work and migration will leave too few people in the countryside to farm enough produce to ensure the food security of rural and urban net food buyers.

Rice production in Vietnam is under pressure from loss of paddies to more profitable products such as fruit and fish or conversion to industrial use while the population is growing with a greater appetite for meat which increased demand for feed, some of which is made from rice. Yields, already among the highest in the world at over 5 t/ha, continue to grow during the 2000s but at decreasing rate. Sustaining output will require substantial investment in inputs, technology and infrastructure to counter the problems of diminishing marginal product. Vietnam should maintain a positive rice balance until 2020 but beyond could prove challenging, particularly as climate change intensifies (Nguyen 2010, pp. 11–17). If states of the wider Mekong region are unable to reserve sufficient land for producing food then the alternative may be greater dependence on imports or higher domestic prices to ensure food production remains competitive for land and water relative to other crops or activities. Net food buyers will be squeezed if their incomes fail to rise faster than food prices.

Agriculture in the wider Mekong region has to accommodate the uncertainties of climate change, which may intensify beyond 2030, and the demands of a population projected to rise from 275 million in 2009 to around 320 million in 2030, which will be more urban and demand richer foods requiring more resources (Johnston et al. 2010, pp. 9–10, 13).

2.2 *Food Resources*

Food in the wider Mekong region is the product of the agro-ecological tapestry of coastal lands and waters, deltas, the Tonle Sap, lowland plains and plateaux, and farmed and forested uplands. Almost half the region's rice is cultivated in the deltas, which also yield a rich supply of farmed and wild fish. The coasts not surprisingly are also a major source of fish. However, varied terrain and resources gives rise to a surprisingly diverse range of products including rice, vegetables, pulses, fruit, livestock and industrial materials such as rubber. The lowland plains and plateaux are increasingly the focus of plantations, which compete with traditional agriculture and needs for land and water. Plantations are also intruding into the fragile uplands, which account for about half of the wider Mekong region, where environmental change is considerable and farming in transition as traditional shifting cultivation is replaced by settled or semi-industrial practices. Cash crops, such as coffee, tobacco, tea and rubber, are common (Johnston et al. 2010, pp. 5–9).

The wider Mekong region Environment Outlook observed between 1992 and 2002 land arable use or forests expanded while crop land per capita increased in Yunnan and Vietnam, meanwhile crop land per capita and forests decreased in Cambodia, Lao PDR, Myanmar and Thailand (Johnston et al. 2010, p. 2). Land degradation affected between 10% and 40% of land in the countries of the wider Mekong region, with Thailand, Vietnam and China's Yunnan the most severely affected due primarily to forest loss and agricultural intensification. Valuable ecosystem services have as a result been damaged (Johnston et al. 2010, pp. 3–4).

3 Consequences of Economic Development for Food Security

3.1 *Dams on the Mekong Mainstream*

An ambitious programme for developing hydroelectric dams is underway in the Mekong River Basin. Dams, especially those proposed for the lower mainstream, are strongly supported by governments which argue there is no alternative for securing energy and financing public spending. However, scientists, civil society organizations and some governments have either expressed considerable concern or outright opposition because the environmental changes are difficult to determine and raise the prospect of major impacts on ecosystem services, agriculture and aquaculture, and livelihood sustainability and resilience in the Mekong basin and beyond. As a result of the acute interest dams are among the most studied developments in the wider Mekong region. In 2010 the Mekong River Commission, which advises the lower four Mekong states on river management, undertook a strategic environmental assessment of dams proposed for the lower mainstream of the Mekong.

The Mekong as a free-flowing river ecosystem is the world's largest freshwater fishery and the most biodiverse in the world after the Amazon. An oft-cited, but hard to substantiate, Figure claims 65 million people in the lower Mekong depend upon the free-flowing river for their protein. The strategic environmental assessment reported people in the lower four Mekong countries lead the world in freshwater fish consumption, with Cambodians eating 8.6 times the world average. In addition to freshwater fish the river's nutrients, flood pulse and habitats help support the rich coastal fisheries of the Mekong delta. Seasonal floods also refresh and revitalize fields, ponds and canals of the Cambodian flood plain and the Mekong delta. The agriculture and aquaculture of these lands secures the food of Cambodia and Vietnam, and moreover through international trade makes a significant contribution to local livelihood security and global food security. Furthermore, current output may be below the sustainable production frontier because investment has been insufficient to research potential for improving methods and supplying additional resources.

Large dams are generally net negative for ecology, economy and society (World Commission on Dams 2000, pp. 68, 92–93, 130). The Mekong River Commission's strategic environmental assessment of mainstream dams notes extracting kinetic energy for power generation will reduce energy flow within the river which powers hydro-ecological and geo-morphological dynamics of river's ecosystems (ICEM 2010, p. 70) and severely curtail biologically-important transition seasons (ICEM 2010, p. 74). The flood pulse and silt loads of the river replenish the Tonle Sap, Cambodia's rice and fish basket, and support the delta against the erosion and salinity of the sea (Hori 2000, pp. 20–21) will be upset, distorted and obstructed by dams.

Modelling indicates complex impacts. Flows will change and there will be less silt. Of the remaining silt more will settle in the Bassac arm of the Mekong from Cambodia into Vietnam which will exacerbate floods. With less silt reaching the delta more land and wetlands will be lost to erosion (Le et al. 2007, pp. 115–116). Flows of sediments and nutrients for primary production will fall sharply in the lower basin as well as in coastal waters, which currently support a large fishing industry (ICEM 2010, pp. 69, 79).

The overall effect of the mainstream dams will be to replace the free-flowing river with a series of large reservoirs. Such a transformation will compound the effects, which in themselves are projected to be even greater than mainstream dams

Table 3.5 Conservative impacts of mainstream and tributary dams in 2030 (baseline 2000)

Mekong dam distribution	Total sediment & nutrient decline in Cambodia & Mekong delta	Fishery catch reduction tonnes/year	Watershed lost for fish migration	Value of impact on fisheries/year	Riverbank agricultural losses/year
Upper mainstream & lower basin tributary	50%	0.21–0.56 m	37.3%		
Lower-basin mainstream	75%	0.55–0.88	81%	$476 m	$50 m

Source (ICEM 2010, pp. 13, 17, 63, 95–96, 99, 101, 103)

if present plans persist (Ziv et al. 2012), of dozens of dams under development or planned for lower Mekong tributaries, primarily in Lao PDR, as well as the mainstream dams on the Mekong in China (Table 3.5).

Conservative estimates indicate the fall in fish catch caused by mainstream dams will exceed the total livestock production of Cambodia and Lao PDR. Reservoir fisheries might at best offset 10% of the losses caused by dams (ICEM 2010, pp. 63, 95–96, 99, 101, 103). The food and livelihood security of tens of millions of people who depend upon the free-flowing Mekong is in question (Chungyalpa 2010, pp. 2–4). Moreover, the hundreds of species of migratory fish and their place in the ecosystem have not been adequately studied (Baran 2006, pp. 7–8, 33–36) to understand their potential for aquaculture and the impacts of dams.

Dimensions estimate 2030 – mainstream dams

Dimension	Impact
Availability	▪ Sharp reduction in fisheries with unpredictable secondary impacts across the ecosystem which may have consequences for food production ▪ Reduction in water quality and nutrients crucial to agriculture and aquaculture plus altered distribution due to changes in flood pulse and cycle ▪ Reduction in local production potential ▪ Reduction in option value for investing in developing ecological infrastructure for food production
Accessibility	▪ Reduction in the agency of farmers and fisherfolk because control over water quality, flow, flood pulse, sediment transport and nutrient load will be captured by dams ▪ Livelihoods of food producers and associated food independence will be impaired or eliminated ▪ Those affected will need more money to purchase agricultural inputs to compensate for reductions in ecological quality of the river or to buy food ▪ Labour supply and migration to cities likely to increase

(continued)

Dimensions estimate 2030 – mainstream dams	
Dimension	Impact
	▪ Incomes may not rise sufficiently to secure food affordability from markets ▪ Food security will weaken further for marginal food consumers
Utilization	▪ Impaired livelihoods and migration may leave people more exposed to poor sanitation and requiring additional knowledge to access and use unfamiliar foods from markets
Stability	▪ People will be exposed to a reduction in the sustainability and resilience of food supply because of declines in diversity and greater dependence on regional and global food trade at a time of increased stress due to rising world population, economic development and climate change ▪ Vulnerability to acute shocks likely to increase if incomes do not rise sufficiently to ensure food affordability

Enabling factors estimate 2030 – mainstream dams	
Enabling factor	Impact
Demography	▪ Impairment of agriculture and aquaculture will reduce livelihood opportunities driving migration to cities and generating additional pressures challenging household viability ▪ Agricultural labour force and skill base may contract ▪ Migration of people with limited skills for sustainable urban livelihoods will contribute to higher slum population, increased labour supply especially in informal employment, and vulnerability to risky behaviours
Resources	▪ Availability and quality of water will change undermining the ecosystem patterns and flows and some agricultural land will be lost ▪ Moderation of flood cycles and reductions in sediment flow may leave the Mekong delta more exposed to coastal erosion, seawater intrusion and salinisation
Infrastructure	▪ Ecological – major shifts with hard to predict consequences for nature and distribution of ecosystem services and flows for food production ▪ Hard – irrigation may expand supporting expansion of off-season food production but with risks of salinization, electricity supply may increase supporting more food processing and refrigeration but affordability will be an issue (however alternative sources of electricity are feasible and could be developed subject to policy) ▪ Soft – people will require skills and knowledge appropriate for the changes in ecosystems and food security
Markets	▪ Existing equilibria of food and informal or low-skill labour markets will be upset by increasing demand for food and expanding supply of labour ▪ Declines in fish production will disturb protein markets forcing major changes which could *inter alia* see production of alternatives, such as livestock, increase generating additional environmental and social impacts

(continued)

Enabling factors estimate 2030 – mainstream dams	
Enabling factor	Impact
Policy	▪ Incentives and regulation will require revision to expand alternative and additional food sources ▪ Urban development will have to accommodate rising migration and additional food security challenges ▪ More resources may be required for emergency food stocks

Assessment	
Negative	▪ Sharp declines in fisheries expected by 2030 ▪ Impacts cannot be mitigated, likely irreversible ▪ High risks for food production in Mekong delta, Cambodia ▪ Large-scale impact on protein supply ▪ Ecosystem transformed, reduction in primary productivity

3.2 *Large-Scale Mekong Water Diversion: Thailand and Lao PDR*

Diversion of water on a large scale from the Mekong mainstream or tributaries in Lao PDR to expand irrigation in Thailand, particularly in the semi-arid northeast, has been identified as the solution to increasing agricultural output, including food, and improving livelihoods. Prospects for food production, and improvements in livelihood security, will however be tempered by local impacts and regional impacts as a result of externalities which will be experienced downstream, especially in the Mekong delta, with consequences for food security. Full implementation of diversion plans would require engineering pipelines, aqueducts, reservoirs and pumping stations. Such works may take several years of planning and construction, perhaps a decade. The effects would thereafter play out over several more years, becoming clearer around 2030.

Several studies and inquiries have concluded diverting water from the Mekong and its tributaries to supply greater irrigation in northeastern Thailand to alleviate aridity and salinity, which currently constrain agriculture, would exacerbate and expand salinity. This proved to be the case with the Rasi Salai dam and irrigation project. Thus an expansion of irrigation is self-defeating giving the nature of the geophysical environment because greater salinity would overwhelm initial gains in cultivation. Expanding irrigation to increase cultivation also raises the prospect of greater use of agro-chemicals detrimental to the ecosystem which will flow back into the Mekong (Molle and Floch 2008, p. 202).

Irrigation's assumed efficiency is not well proven especially given limitations in labour, administration and markets. The efficacy of small and medium irrigation projects, including wells, weirs and reservoirs, developed in Thailand has not been comprehensively assessed (Molle and Floch 2008, p. 203). Less than a sixth of the land under irrigation department surface water irrigation in schemes in northeastern

Thailand was being planted in the early 2000s. Barriers to production included labour shortages and farmers' preferences for planting small areas with high value crops, like chilli and watermelon, instead of the whole farm with rice (Nesbitt et al. 2003, p. 93). Large-scale water management has historically been more beneficial to agri-business than to smallholder farmers in Thailand, has concluded Lertsak Khamkhongsak of the Ecological and Cultural Change Studies Group in Udon Thani (Winwong 2008).

Thus, livelihoods may not improve as expected constraining improvements in the affordability of food. Moreover, large agribusiness enterprises and markets will require careful regulation to prevent abuse of market power while also ensuring sufficient returns required for commercial viability of large-scale chemical- and petro-intensive agribusiness. Even if irrigation does facilitate sustained expansion of production by smallholders or firms they will cultivate whatever products provide the greatest returns. While that is in their best interest the overall result may not be in the best interest of society if industrial or export-oriented crops are favoured because of higher profits.

Additional diversion for irrigation of large parts of northeastern Thailand is likely to exacerbate water stress in the Mekong. Although agricultural withdrawal of water may be less than 4% of annual flow, agricultural peak demand coincides with the dry season when flows are at less than 10% of total annual flow (Nesbitt et al. 2003, p. 87). In every other year of any given period water consumption equals 40% of dry season flow in the Mekong (Nesbitt et al. 2003, p. 102). Water diversion upstream will exacerbate low water levels during the dry season in downstream areas such as the Mekong delta with detrimental consequences for rice production, fish, mangroves and saline intrusion (Le 2001, pp. 2, 7); (Nesbitt et al. 2003, pp. 101–102).

The impacts of diversion of water from the Mekong in Lao PDR to northeastern Thailand may not occur in isolation. Proposals and plans have been presented in all riparian states to rehabilitate, expand or develop irrigation raising demand for water diversion around the Mekong basin (Molle 2005, pp. 2–7). Less than a sixth of the lower Mekong watershed suitable for lowland agriculture is reached by some form of irrigation scheme. While this might suggest considerable potential for expansion, the poor quality of soils in many areas will require investments to improve productive potential in addition to any irrigation (Nesbitt et al. 2003, p. 87). Such challenges and limitations of the efficiency and effectiveness of irrigation have already been experienced in northeastern Thailand.

Dimensions impact estimate 2030 – Mekong diversion

Dimension	Impact
Availability	▪ Expansion of production capacity does not *ceteris paribus* imply an increase in food production or an increase in the production of food most needed in the wider Mekong region ▪ Expansion of production scale because of irrigation in northeast Thailand may alter production economies to favour industrial crops over food crops ▪ Potential, but not assured, short-term increase in production of some food in irrigated areas

(continued)

Dimensions impact estimate 2030 – Mekong diversion	
Dimension	Impact
	▪ Reduction in dry season water flow due to irrigation diversion through the Mekong system will raise costs and/or reduce output downstream for a range of crops ▪ Long-term increase in production threatened by salinity in irrigated areas plus greater saline intrusion in the Mekong delta due to reduction dry-season flow because of upstream water diversion
Accessibility	▪ Control of irrigation may not be vested with farmers ▪ Minimum flows guaranteed by Mekong Agreement may not be enforceable or prove to be sufficient to secure production downstream raising issues of control ▪ Diversion may encounter conflicts with hydropower demands ▪ Rural livelihoods and thus food affordability may not improve if incomes of farmers and labourers do not sustainably increase faster than food costs
Utilization	▪ Greater salinity caused by irrigation could pose risks for groundwater used for drinking and cooking ▪ Reductions in dry-season river flows due to diversion may exacerbate drinking water supply problems in the Mekong delta
Stability	▪ Climate variation is expected to affect seasonal distribution of precipitation by 2030 (Eastham et al. 2008, pp. 19–27) which raise uncertainties over transport and storage capacities of diversion and irrigation ▪ Acute or chronic drought will imperil water available for diversion especially given demands of hydropower ▪ If diversion rates are maintained upstream during drought dry-season flows downstream will be further reduced ▪ Water stress will severely test existing non-binding agreements guiding the use of the Mekong by the lower four riparian states ▪ Long-term output in question because of salinity issues in northeast Thailand and the Mekong delta which diversion for irrigation from Lao PDR will exacerbate

Enabling factors estimate 2030 – Mekong diversion	
Enabling factor	Impact
Demography	▪ Increase demand for labour in northeastern Thailand ▪ Reduce flow downstream may reduce incomes in agriculture and aquaculture driving migration to cities
Resources	▪ Adjustment to water flow patterns and balances in the Mekong system affecting hydrology and ecosystem services at up to basin scale ▪ Changes in land use, primarily agricultural expansion and intensification, may favour increased production of industrial crops at the expense of food production in northeastern Thailand
Infrastructure	▪ Ecological – distorted by salinization in northeastern Thailand and the Mekong delta causing adjustments in the type and flow of ecosystem services ▪ Ecological – decline should production follow large-scale chemical- and petro-driven monocrop models which may cause impacts beyond the area of use

(continued)

Enabling factors estimate 2030 – Mekong diversion	
Enabling factor	Impact
	▪ Hard – planning challenges because of the uncertainties of climate changes and bargaining over water rights plus will require budgets and manpower for ongoing maintenance and adjustment ▪ Soft – requires new knowledge and retraining to deal with opportunities and impacts brought by changes in water flow and distribution
Markets	▪ Increased production capacity will require additional labour causing wages to rise if labour supply proves relatively inelastic because of population distribution or controls on migration ▪ Excess output of food products already in abundance may depress prices with viability effects for livelihoods of less competitive smallholders across the wider Mekong region
Policy	▪ Zoning and production will require management through regulation and incentives to balance output against the requirements of strong food security nationally and regionally ▪ Control, participation, management and costs of diversion and irrigation require nuanced policy to prevent capture by interest groups at the expense of society ▪ Ecological impacts such as salinization and pollution from agro-chemicals require regulation for prevention ▪ Diversion and irrigation requires trans-national policy coordination and design to avoid trans-basin externalities
Assessment	
Negative	▪ High risk of salinization ▪ Production environment already marginal ▪ Labour may not be available to increase production ▪ Gains in production may not be sustainable ▪ Diversion risks worsening low water flow in dry season ▪ Diversion will compound effects of mainstream dams ▪ Diversion may be in conflict with mainstream dams ▪ Transnational impacts especially on Mekong delta

3.3 Mitigation and Adaption to Rising Sea Levels in the Mekong Delta

The level of the sea is rising faster than would otherwise be the case because of climate change due to emissions of carbon dioxide and other greenhouse gases. Rising seas pose a great challenge to the methods of production and the composition of food obtained from the coasts of the wider Mekong region. The Mekong delta[6] is considered as a case because it may be particularly vulnerable, with impacts amplified by the impacts of mainstream dams on the Mekong upstream.

[6]Extending over 36,630 km^2 in Vietnam and 12,870 km^2 in Cambodia (Le et al. 2007, p. 110)

The Mekong delta plays a key role in the food security of Vietnam and the world. Fisheries yield in 2010 of 2.9 million tonnes, or 57.2% of Vietnam's catch, was almost three times greater than at the turn of the century. Farmed aquatic products amounted to 1.9 million tonnes, equal to 71.7% of national output, and six times greater than a decade earlier. In 2010 the rice harvests totalled some 21 million tonnes, equivalent to 53.9% of national production. In 2011 Vietnam's rice exports were 7.1 million tonnes, worth $3.507 billion (VFA 2012), of which 90% was produced in the delta equivalent to 23.22% of world trade in 2010[7] or approximately the combined demand of the top rice importers the Philippines, Nigeria, Iran and Saudi Arabia (Thukral 2011). The national food and well-being security of Vietnam, and elsewhere, is in no small part guaranteed by the vitality of the Mekong delta.[8]

Production from the Mekong delta however faces rising seas caused by climate change. A 1 m rise in sea-level would leave 90% of the Mekong delta vulnerable to inundation (Rowe 2011). The extent and speed of the rise in sea level is difficult to specify with great confidence, nevertheless research is pointing towards the sea rising further, faster. A sense of the scale and uncertainty of change is therefore necessary for estimates of food security in 2030.

The Intergovernmental Panel on Climate Change projects sea level to rise up to 0.59 m by 2,100 under the highest emission scenario, which does not consider impact from net reduction of global ice stock into water (IPCC 2007). However, research since then gives reason to treat the projection with caution and contemplate accelerating net reduction of ice with implications for the efficacy of engineering in the delta to counter floods and saline intrusion.

The IPCC projection for sea-level rise will be exceeded if the ice sheets of Greenland and Antarctica continue to lose mass at the current rate three times which is three times that of mountain glaciers and ice caps (Rignot et al. 2011). Paleoclimate analysis of the last million years indicates strong amplifying polar feedbacks from moderate warming of less than 2°C, all but locked in by the present stock of greenhouse gases in the atmosphere, which would leave the seas several metres higher than today by 2,100. If the analysis is correct then atmospheric carbon dioxide concentrations, currently around 390 ppm, must stop rising and rapidly return to 350 ppm (Hansen and Sato 2011, pp. 25–27).

Emissions and warming trends remain strong however because policy and rhetoric assumes a 2° increase in mean surface temperatures is the threshold for dangerous climate change instead of extremely dangerous climate change. Moreover, a 2° increase is certain because current emission trends are well advanced risking 3–4° of warming absent dramatic change in policy, including planned carbon austerity in developed countries and rapid shifts away from fossil fuels in other countries by 2020 (Anderson and Bows 2011, pp. 38–42).

Research projects by the British government (Lowe et al. 2009, pp. 31–33), the Proudman Oceanographic Laboratory (New Scientist 2008), the Arctic Monitoring

[7] International Rice Research Institute http://ricestat.irri.org/vis/wrs_quickCharts.php

[8] Production data from the General Statistics Office of Vietnam unless stated otherwise

and Assessment Programme (Doyle 2011), the Sea Level Research Laboratory (Collins 2011), and the Niels Bohr Institute (2011) incorporating contracting ice sheets and realistic to pessimistic assumptions of efforts to reduce emissions forecast rise in sea level of 0.5–1.9 m by 2,100, with the upper boundary still considered unlikely but not impossible. The assumptions and evidence are such that they are still subject to considerable debate, evident in the range for sea level rise, which requires research on the great ice caps, their melting and contribution to rising seas (Gregory and Lowe 2010).

Notwithstanding the uncertainty and research gaps research implies an increasing likelihood of a 1 m rise in sea level by 2,100. The Ministry of Natural Resources and Environment of Vietnam published a projection in 2009 of sea level rise reaching 0.3 m between 2030 and 2050 (Toan et al. 2011, p. 9). The foregoing review of research however suggests sea level 0.3 m higher in 2030 than 2000 assuming for simplicity a linear rate of change, although because of forcings and phase-change effects in reality the rate may be non-linear. Higher tides will be exacerbated by more frequent and powerful expected as a result of climate change exposing more of the delta more of the time to greater damage. Higher tides and worse storms will compound the saline intrusion and floods farmers in the delta have been working with for centuries.

Saline intrusion is a growing problem peaking late in the dry season around April and May falling to a nadir at the end of the monsoon around October. The Ministry of Agriculture and Rural Development estimates out of 650,000 ha where high-yield rice is grown in the lower Mekong delta about 100,000 faces high risk of dry-season saline intrusion. Reductions in yield due to salinity vary according to time and duration of exposure during the life cycle of rice (Nhan et al. 2012, pp. 1–2, 5).

The full impact of saline intrusion on output is not clear. Currently a few 100,000 t of output (Nhan et al. 2012, p. 13) appear at risk and 273,000 ha are threatened (Toan et al. 2011, p. 17). Extending rice cultivation in the upper Mekong delta is expected to increase saline intrusion in coastal lands causing rice output falling 120,000 ha (Toan et al. 2011, p. 22). It is not clear if the reduction in output will be outweighed by expansion of production in the upper delta. The effect on saline intrusion of a slight increase in water flow in the Mekong during the dry season by 2030 is unknown (Eastham et al. 2008, p. 62). Nevertheless, as seas rise more land becomes vulnerable to saline intrusion or saline residue from storm surges suggesting greater difficulties, uncertainties and costs in sustaining rice production and farmer livelihoods in coastal lands.

Mitigation focuses on building hydrological-management infrastructure, adapting farming techniques and models, including more shrimp and saline-tolerant rice, enhancing human skill, and adjusting policy and institutions to explicitly recognise the problem (Nhan et al. 2012, pp. 2, 9). Around 100,000–200,000 t of rice output could be maintained by switching to saline-tolerant varieties plus agrochemicals and fertilizers without resort to investing in large-scale hydrological infrastructure. The effect of these measures would be further enhanced by better weather forecasts, adjusting cropping seasons, revising rice-transplantation practise, and improved desalinization of paddy soil by small scale irrigation systems (Nhan et al. 2012, p. 13).

Four major sluice gates could prevent saline increases across 273,000 ha and increase rice output by 160,000–300,000 t. Enclosing parts of the Mekong delta with major sluices risks serious consequences including increasing acidity leaching from extensive acid-sulphate soils and concentration of domestic and industrial water pollution (Toan et al. 2011, pp. 17, 22).

Similar negative consequences are experienced with other great works, such as extensive embankments and dykes, which have been expanding to control floods. Fields are also less fertile because they are no longer flooded by nutrient-rich waters. Meanwhile, water is displaced and concentrated into unprotected lands which suffer worse floods. Moreover, high embankments are at greater risk of catastrophic failure because embankments accelerate water flow (Le et al. 2007, p. 116).

In 2030 flooding in the Mekong delta will be more frequent and severe. The median flooded area of the Mekong delta in 2030 relative to 1951–2000 will be approximately 10–20% larger with projections spanning a wide range. Notably, the lower boundary of the flood range for 2030 lies above the historical trend. Higher minimum dry season flows expected in 2030 will have to counter the rise in sea level to reduce saline intrusion (Eastham et al. 2008, p. 66).

However, the development of dozens of dams upstream, particularly in Lao PDR, will reduce flood peaks and may lower floods in the Mekong delta. But dams will also lead to siltation of the Mekong estuaries which will cause flooding in excess of current levels in as little as 30 years (Le et al. 2007, p. 112). If dams are managed to ensure sufficient minimum flow during the dry season saline intrusion may be moderated, which could increase rice output by 300,000 t. However, if the requirements of power generation are prioritized over minimum flows saline intrusion may expand especially during droughts, which are expected every 7 years, putting one million tonnes of rice at risk (Toan et al. 2011, pp. 5, 13–15, 22–23).

The environment of the Mekong delta has changed considerably over the last three decades because of efforts to increase rice production, including new varieties and techniques and flood control structures amid economic transformation. Great increases in rice output due to these measures have however been accompanied by new challenges and risks. Agro-chemical and fertilizer run-off pollution has increased, floods have worsened in some places, ecosystems and biodiversity suffer, and water quality has declined. In the upper Mekong rice intensification has substantially raised dry-season water demand. In addition to local effects, common-pool fisheries have been damaged (Käkönen 2008, pp. 209–210).

Furthermore, the losses, except during drought years, are insignificant for regional food security given rice production in the Mekong delta of 21 million tonnes in 2010. However, they may be significant for world trade if losses reduce quantity available for export. A shift in emphasis from spending on hard control engineering to proven and promising adaptation, such as saline-tolerant rice, more rice-shrimp rotation, and focusing on growing rice in the mid- and upper delta (Nhan et al. 2012, p. 13) could be a more promising cost-benefit trade-off for food security. Nevertheless, the current focus on hard infrastructure for flood and saline mitigation implies higher expenditures and potentially less adaptation investment, and likely resilience outcomes, than might otherwise be the case.

The negative impacts of sea-level rise and the policy response on the high-level of productivity may be exacerbated by the effects of climate change on aquaculture and fisheries. Experiments indicate fish larvae of an estuarine species commonly used for environmental risk assessments are highly susceptible to higher ocean acidity levels, already occasionally observed in some waters, likely to be common by 2,100 due to rising carbon dioxide concentrations in the atmosphere (Baumann et al. 2011). Not dissimilar results have also been found for Atlantic cod (Frommel et al. 2011). Furthermore, climate change also affects the supply of fish by altering their distribution and migration, reducing size and lifespan, and damaging food stocks of shellfish and seaweed through acidification. Shifts and reductions in stocks will upset the economics and livelihoods of tens of millions of people (Sumaila et al. 2011, pp. 449–451, 454). Such effects imply prospects for aquaculture and fisheries offsetting losses in coastal rice output and livelihoods may be limited.

In conclusion, the Mekong delta presents a complex challenge for food security. Saline intrusion and flooding will intensify under increasing pressure from the vice of rising seas and modified river flow. Furthermore, efforts to control and mitigate saline intrusion and flooding using dykes, sluices and other structures damage environmental quality, burden budgets and risk catastrophic failure in exchange for what may turn out to be limited gains in rice production. Moreover, the long-term cost of maintaining and enhancing such structures in the face of seas rising by a metre or more come 2,100 and beyond may not be sustainable. However, current developments in adaptation of crops and techniques, livelihoods and economy, in other words investment in ecological and soft infrastructures, offer a promising sustainable alternative for stronger food security.

Dimensions impact estimate 2030 – rising seas

Dimension	Impact
Availability	▪ Slight reduction in rice output during normal years ▪ Significant reduction in rice output during drought or extreme weather ▪ Increasing risk of losses beyond 2030 ▪ Impact of losses may be compounded by losses elsewhere in vulnerable areas of Vietnam especially coasts and Red River delta ▪ Losses could exacerbate challenge of maintaining rice balance after 2020[9] ▪ Export volumes will fall if a domestic food security first policy can be enforced ▪ Quality of rice, fish and other products may be tainted by residues of agro-chemicals, ▪ fertilizer and other pollution generated by intensification or hydrological engineering

(continued)

[9] See 2.1. (Nguyen 2010, pp. 11–17)

Dimensions impact estimate 2030 – rising seas	
Dimension	Impact
Accessibility	▪ Households suffering production losses will suffer weak food security if they lack stocks or cash for market purchases ▪ Market-dependent households in rice-deficit countries might experience higher prices if Mekong output falls enough to significantly reduce exports ▪ Control of water for soil fertilization and flushing of acids shifting from natural flood flows to agencies designing and managing hard infrastructure
Utilization	▪ Potable water used for drinking and cooking rice increasingly may face growing risks of pollution from saline intrusion and floods ▪ Agro-chemicals, fertilizer and acidification may be detrimental to quality of potable water
Stability	▪ Climate change will cause more intense and more frequent acute shocks with less time for recovery ▪ Chronic deterioration from saline intrusion, floods and pollution may reduce tolerance for acute shocks ▪ Resilience will face greater challenges until sea level and climate stabilise ▪ Effects of upstream dams compound climate change pressure on resilience ▪ Switching emphasis from adaptation through hard infrastructure to soft and ecological ▪ infrastructure offers a more sustainable approach to long-term reduction in climate ▪ change impacts

Enabling factors impact estimate 2030 – rising seas	
Enabling factor	Impact
Demography	▪ Climate change impacts may be another driver for migration from agriculture to industrial urban society ▪ Rising production costs or charges for hard infrastructure could force marginal farmers to exit ▪ Vulnerability of weak food security households could be magnified by climate change impacts if rice balance deteriorates after 2020
Resources	▪ Freshwater environment will contract replaced by more brackish and saline environments ▪ Soil acidification may worsen ▪ Less land may be available ▪ Area for saline aquaculture will expand ▪ Quality of land and water could be reduced by agricultural and other pollution
Infrastructure	▪ Ecological – in transition towards a more saline environment ▪ Ecological – damage from agricultural pollution and effects of hard infrastructure on water flow and quality ▪ Hard – expansion of sluices, dykes and other structures affects water availability and distribution while introducing new risks and pollution ▪ Soft – farmers, officials and investors require new knowledge and skills to adapt agriculture to rising seas

(continued)

Enabling factors impact estimate 2030 – rising seas	
Enabling factor	**Impact**
Markets	▪ Rising output of shrimp rotating or replacing rice may depress prices threatening returns and viability for farmer livelihoods ▪ Costs, especially if agro-chemical intensification expanded, may rise risking higher prices undermining food affordability ▪ Markets for ecological services may support ecological infrastructure investments in place of current emphasis on hard infrastructure
Policy	▪ Manage trade-offs of net short and long-term benefits of hard infrastructure versus ecological and soft infrastructure ▪ Long-term planning and preparation for migration of households and agriculture to higher ground, which may require years of ecological modification for agriculture
Assessment	
Negative	▪ Sea level rise may be faster and higher than expected ▪ Saline intrusion will increase ▪ Mekong coastal rice production will be hard to sustain ▪ Coastal production at risk elsewhere in Vietnam ▪ Hard engineering solutions are costly and ineffective ▪ Adaptation is more sustainable for securing food production ▪ Rice exports may decline to zero by 2030 partly due to rising seas

3.4 Industrial Crops: The Case of Rubber

Expansion of industrial crops, such as rubber and pulp, weakens food security by reducing land available to produce food, expanding the pool of net food buyers, and damaging the ecosystem services which operate at small and large scale to support food production. The scale and speed of change is hard to miss on the ground in Luang Nam Tha and parts of the Bolaven plateau in Champasak and other provinces. Plantations for rubber have replaced farms and commons such as forests (Fujita 2007, p. 2.ii), for example rice in Luang Nam Tha (Shi 2008, p. 32) and forests and crop lands in Champasak (Baird 2010, pp. 14–19). Plantations can be particularly hard for poor households (Fujita 2007, p. 22), which often cannot afford to invest in rubber (Sithong and Thoumthone 2006, p. 121), and rural communities in general (Rigg 2006, p. 126) because foods foraged in forests lost are associated with better nutrition (Galland 2007, p. 6) and up to half of cash income (Foppes and Ketphanh 2007).[10]

Plantations in Lao PDR weaken access and availability of in several ways. One, where plantations replace or impede access to land used for field crops or forests rich in natural foods people turn to markets (Baird 2010, p. 22). Two, remaining forests and other commons are subject to demand from more people which exceeds

[10] Cited by (Barney 2007, p. 49)

the limits of sustainable yields (Department of Forestry 2007, p. 8). Three, loss of forests compounds problems facing farmers struggling to maintain the fallow periods for sustainable swidden agriculture (Baird and Shoemaker 2007, p. 868) who have been caught up in aggressive efforts to eliminate unsustainable pioneer slash-and-burn agriculture by provincial authorities apparently misinterpreting government policy (SWGUD 2008, p. 7). Four, income from producing rubber or working on plantations may be erratic or insufficient to ensure people can afford sharp increases in the necessity to buy most of their food from markets. Five, migration for work or settlement elsewhere is a likely response for households, especially in growing communities, facing inadequate incomes and difficulty accessing food because of the effects of plantations.

Introducing large-scale plantations can also be hard for ecosystems to absorb and adapt too. The scale of disturbance and destruction wrought by the change, along with secondary effects such as increasing foraging pressure on what forests remain, in Lao PDR is overwhelming the capability of ecosystems to adapt and regenerate (Galland 2007, p. 4). As a result heterogeneous landscapes of forests and traditional agriculture are becoming homogenous expanses of rubber or other industrial crops such as acacia or eucalyptus for pulp.

After development plantations can have effects which reach well beyond their boundaries. In addition to upsetting ecological services crucial to food production, such as pollination or keeping pests in balance, agro-chemicals leach into soil and water flowing through the ecosystem (SWGUD 2008, pp. 1–2) raising the prospect of contamination of food production. Rubber and other Industrial plantation trees also compete for ground and surface water. Xishuangbanna, a prefecture of China's Yunnan bordering Lao PDR, replaced forests with rubber plantations on a large scale. In places streams and wells have run dry because of the demands of rubber trees (Shi 2008, p. 60); (Qiu 2009, p. 246). Furthermore, evaporation and temperature have also increased. The damage caused to ecosystem services by rubber is detrimental to the interests and sustainability of the wider economy (Qiu 2009, p. 247). The evidence from Yunnan echoes findings of strong demand exerted on soil hydrology by fast-growing plantation trees in varied contexts studies elsewhere have found (Bruijnzeel 2004, pp. 198–199).

The new landscapes emerging from the plantation transformation promise to change ecological services which underpin the sustainability and resilience of food production. New knowledge and techniques may be required. Farmers may have to use more agro-chemicals to compensate for the reduction or loss of ecological services that were obtained from heterogeneous landscapes. Such shifts in production imply higher costs and greater dependence on suppliers of inputs. On the other hand, the costs of agro-chemicals and commercial seeds have been associated with higher output, although sustaining such yields without causing further damage to the ecosystem appears to be difficult.[11] If yields and incomes of smallholder farming

[11] As an example a suspected case was uncovered and analysed in Bokeo (Fullbrook 2011, p. 33)

cannot be maintained in the face of the realities of plantation landscapes will more people leave their place in the land taken by more plantations?

Despite the considerable ecological, economic and social impacts for food security presented by rubber, and other tree plantations, expansion has been rapid, probably out pacing research and understanding of the long-term intra- and inter-basin impacts. Plantations generate several revenue streams for the government, including concession fees and taxes. Plantations may also provide rents which corrupt policy development, implementation and enforcement. Baird (2010), Shi (2008, p. 31) and the government's Sub-working Group on Uplands Development (2008, p. 3) describe instances of local or provincial civil and military officials benefiting from ensuring investors privileged access to the land for rubber.

Expansion of plantations also coincides with the government's expressed intentions to reverse deforestation due to legal and illegal logging. Stated goals have included increasing forest cover from 9 million hectares (42% of land) to 12 million hectares (53%) by 2010 (Department of Forestry 2007, pp. 3–4), with 15 million hectares (70%) under tree cover by 2020 (SWGUD 2008, p. 2). Plantations of nearly two million hectares have been proposed by investors (Department of Forestry 2007, p. 3).

Reforestation however does not automatically restore the environment, particularly ecological flows across the landscape which support food production, for public well-being. The effects of cutting or planting trees at large scale on a landscape can vary considerably over time and space according to the interactions of many factors including among others soil types and conditions, geology, albedo and evapotranspiration (Bruijnzeel 2004, pp. 187–201, 212–214). In particular dry-season water flows may decline, especially if commercial trees are planted (Calder 2003, pp. 2–4), or might increase after years even decades if native species are planted on soil not fundamentally altered by deforestation (Bruijnzeel 2004, pp. 208–209, 217–218).

Reforestation trees at such scale and speed can only be achieved in Lao PDR by plantations of industrial trees financed by investors because the costs and resources of trying to create something like a natural tropical forest, if indeed that is possible, are beyond the capacity of one of the world's poorest states. What little data is available suggests, notwithstanding doubts over quality, this is indeed the case. The target for plantations of rubber and pulp was 500,000 ha by 2020 (Barney 2007, p. 10). Reported land concessions totalled over 1 million hectares (Hanssen 2007, p. 6). By 2008 rubber plantations had reached 200,000 ha and the government proposed a cap of 300,000 ha (Vientiane Times 2008).

The interests of investors coincide with those of the government which wants more trees and seeks additional revenues forming a reinforcing feedback loop for rubber despite the consequences for food security. The results harm poor villagers and contradict the government's policy of poverty eradication (Department of Forestry 2007, p. 13). Moreover, the trends and drivers for planting rubber and other industrial trees as plantations in Lao PDR will not restore the ecosystem services, including foods pollination and water balances, obtained from diverse primary or secondary forests.

Rubber, and other plantation trees, compete with food production for land and water. Moreover, large-scale monocrop rubber plantations can nudge the ecosystem into a state which may be less conducive to food production in general or production using existing knowledge and technologies. That plantations appear to take precedence over securing food for strong security is a consequence of powerful financial incentives and policy interests which form a complex unable to accommodate achieving strong food security, if indeed it is even open to recognizing the complexity and interactions of food security. Rubber and other plantations are on trend to continue expansion over the years towards 2030, by which time all opportunities may be fully planted

Dimensions impact estimate 2030 – plantations

Dimension	Impact
Availability	▪ Flow of forest foods continues to decline ▪ Local rice production declines where rubber displaces rice production ▪ Food supply increasingly dependent on external production and markets ▪ Production at large scale vulnerable to declines in ecosystem services caused by expansion of plantations
Accessibility	▪ Smallholders losing control of household land and commons to investors planting rubber ▪ Demands of rubber plantations can reduce or impair farmers access to ecosystem services such as groundwater, pollination and nutrient flows ▪ People losing access to land for producing food or foraging may have insufficient income to secure food from markets
Utilization	▪ Water supplies may be reduced or tainted by the demands or agro-chemical production regimes of rubber plantations ▪ Dietary knowledge may be inappropriate for changes in food composition for people turning to markets to compensate for reduced access to fields or forests
Stability	▪ Food imports and supply vulnerabilities expand to compensate for fall in local food output, if intensification is unsustainable or insufficient, as resources shift to rubber production ▪ Impacts of large-scale rubber plantations threaten long-term shifts in ecosystems and services which shape food production capacity and resilience

Enabling factors impact estimate 2030 – plantations

Enabling factor	Impact
Demography	▪ Rubber is reducing resources for food production when demand is rising due to population growth ▪ Plantation demand for labour exceeds local supply drawing migrant workers increasing pressure on local food resources and markets
Resources	▪ Rubber is in some places replacing forests or fields which supply food ▪ Rubber competes with food production for ground and surface water and nutrients in the ecosystem

(continued)

Enabling factors impact estimate 2030 – plantations	
Enabling factor	**Impact**
Infrastructure	▪ Monocrop plantations like rubber can disturb the structure and function of the ecosystem which supports food production ▪ Rubber agro-chemicals and processing is a pollution threat, especially for water, which will degrade the quality and productivity of the ecosystem for food production ▪ Ecological – water and other flows strained, adjusted or damaged especially in dry season by demands of rubber plantations ▪ Ecological – rubber plantations can damage the diversity, resilience and productivity of ecosystem services supporting food production ▪ Hard – rubber plantations may require new tracks or roads cutting into the ecosystem, increasing water run-off and improving access to remaining commons like forests for poachers and foragers ▪ Soft – food production may require new knowledge, skills and technology to adjust to changes in the ecosystem and water quantity and quality caused by rubber plantations
Markets	▪ Shortages of domestic labour may not increase wages which might improve food affordability because of availability of migrant labour from Vietnam, China and perhaps Myanmar ▪ Plantation labour may increase pressure on food markets especially during times of scarcity ▪ Food markets may become more liquid and resilient due to scale effects as demand increases because of labour demand
Policy	▪ Preference for large-scale rubber and other industrial trees is undervaluing ecosystem services critical for food production resilience and increases ▪ Strong food security is at best a peripheral consideration in large-scale expansion of monocrop rubber and other industrial trees plus forestation in general ▪ The ramifications of rubber, and other industrial trees, as agents of change are not effectively mitigated through management which should maintain the underlying productivity and security of food
Assessment	
Negative	▪ Fundamentally alter and weaken ecosystem function and services ▪ Disrupt water and nutrient flows through the landscape ▪ Replace fields and forests supporting nutrition security ▪ Local and regional impacts from agro-chemicals and processing ▪ Increase demand on local food markets ▪ Increase vulnerability of people and reduce resilience of food production

3.5 *Transportation Infrastructure: The Case of the Yunnan-Cambodia Railway*

Railways expand connections between sources of demand and supply, which may improve the efficiency of the economy by for example reducing the costs and time for transportation. However, while there may be benefits for the well-being of society at large some people may suffer considerable costs, for example if their livelihood is rendered uncompetitive or redundant.

Railways may weaken food security if their construction requisitions land and introduces structures without sufficient concern of the impacts on land, forests and water for food production as well as farmers access. Construction can also be a source of local pollution, particularly of streams and rivers. Construction labour may also temporarily increase demand for food from local markets which if prices rise will harm people with weak food security. Poaching from commons may also increase during construction. Food can be diverted from local or regional markets, particularly during times of scarcity, if richer centres of demand, such as cities or other countries, can exercise their buying power through the conduit of railways and their transportation efficiency. Railways may also reduce costs of economic developments, such as plantations, mining or urbanization, thereby facilitating changes in land use detrimental to food production.

Railways however may also strengthen food security by reducing the costs of importing food quantity and quality of food sufficient for strong food security. Local food production and job opportunities may expand because of the connections made by railways helping improve local livelihoods and incomes which could make food more affordable. If railways help people access stronger labour markets with higher wages then people in formerly remote areas may be in a position to afford strong food security.

In general, railways direct impact on food security will probably be moderate. However, indirectly railways may play a critical role in facilitating developments such as mines which could have a considerable impact on local food security.

Dimensions impact estimate 2030 – railway

Dimension	Impact
Availability	▪ External demand may exert greater pull over local production because of the transport efficiency of railways ▪ External supply may reduce local food costs because railways make imports cost-effective ▪ Land use may change damaging food production
Accessibility	▪ Control and use of resources may move against food production to industrial needs ▪ Labour demand and wages may rise ▪ People may obtain higher wages elsewhere because railways facilitate travel
Utilization	▪ Water pollution during construction may affect food production or domestic supply ▪ Changes in land use facilitated by railways may be detrimental to water quality and availability for domestic use and food production
Stability	▪ Railways may magnify scarcity if controls on food trade are ineffective or may alleviate scarcity by efficiently connecting surplus markets

(continued)

Enabling factors impact estimate 2030 – railway	
Enabling factor	Impact
Demography	▪ Shifts in population density towards stronger labour markets ▪ More possibilities for commuting and migration for work
Resources	▪ Industrial development of natural resources, especially minerals, will be more cost-effective ▪ Bulk agricultural inputs for food production, such as fertilizer, may be more readily available and cheaper
Infrastructure	▪ Ecological – limited direct impacts, except for local water pollution, but considerable indirect impacts ▪ Hard – cheaper, faster transportation, especially of bulk resources and products ▪ Soft – new opportunities facilitated by railways may require specialized knowledge or skills to understand and exploit
Markets	▪ Food, commodities and labour markets will be less fragmented, more cohesive, operating at wider scales ▪ Shocks may be more easily transmitted, including additional demand at times of food scarcity regionally or globally
Policy	▪ Railway regulation will have to balance running costs against affordable passenger fares and operating subsidies ▪ Construction and operation requires careful planning and close supervision to minimise impacts on land and water which may affect food production or domestic water ▪ In times of scarcity railway transport may require close supervision for effective controls on exports to ensure local markets remain well supplied with food
Assessment	
Ambiguous	▪ Limited direct impacts from emissions and effluents ▪ Connectivity is a double-edged sword ▪ Food supply might increase or decrease depending on policy ▪ Access to livelihoods and better wages may improve

3.6 *Mining: The Case of the Bauxite Triangle (Lao PDR, Cambodia, Vietnam)*

The scale of bauxite mining in adjacent areas of Cambodia, Lao PDR and Vietnam, with bauxite deposits around the Bolaven plateau in southeastern Lao PDR alone comparing favourably in quantity and quality to one of the world's largest reserves in the Darling Ranges of Western Australia (Ord 2006a), will change the state of the environment. There is a risk of local and downstream pollution if construction and operations are not closely overseen by well-resourced and highly competent government agencies with sufficient authority to ensure compliance with the highest standards of environmental management. Bauxite mining could weaken food security directly and indirectly during construction and operations which will continue beyond 2030.

Construction of the mines requires clearing land, building processing facilities, storage for tailings, and connections for power, water and transportation. Food production will fall if fields for crops occupy land earmarked for the mine and its facilities. Commons such as grasslands and forests where people forage or graze livestock will also be lost. Usufruct and similar customary rights may also be lost impeding people in their farming and fishing. Such direct impacts occur where population is relatively sparse. While the effects may be insignificant at the regional scale for people in the communities affected the effects will be severe and their mitigation will rest upon compensation programmes which may entail further disruption and upset through relocation.

In addition to such direct impacts, construction of the mine may also have indirect impacts through pollution of ground and surface water which could have transnational effects because of the proximity of borders and the geography of watersheds and river basins. Pollution could taint or kill fish and other food obtained from rivers, tarnish crops through irrigation, and damage domestic water. Resources required for construction, including aggregates, timber and water, may have consequences elsewhere in the region for food production. Many more people over a much larger area, which may pose jurisdictional challenges and raise issues in international law, are at risk from indirect impacts.

Operation of the mines could directly affect food production through chronic air and water pollution if control and containment standards and practises are not mandated or adhered too. Copper and gold mining at Ok Tedi in Papua New Guinea and Grasberg in Indonesia have caused major damage to the environment, especially rivers, with the impacts of the later described as extensive and irreversible (Norway 2006). In Zambia, for example, plants downwind from copper smelters have been killed by 300,000–700,000 t of sulphur dioxide emitted annually (Pegg 2006, p. 378). Zambia, a developing country with decades of experience with the copper mining, has not succeeded in enforcing environmental laws and standards in the face of lapses and irresponsible practises of mining companies (Simutanyi 2008, p. 9). In northern Lao PDR cyanide was spilt at the Pan Australian mine at Phu Bia in 2005 polluting water courses, poisoning fish, causing an acute and possibly chronic impact for food security of hundreds of villages according to state media and other reports (Mineral Policy Institute 2005; Gao 2005). Pan Australian insisted no villagers were poisoned.

Acute pollution is also a risk, particularly from spills of tailings such as the red mud, which contains relatively low concentrations of heavy metals, left over from processing bauxite into alumina. In Hungary a tailings pond dam broke releasing red mud into rivers affecting several countries (BBC 2010) killing fish and damaging crop lands (Redsludge 2010) and leaving a legacy of heavy metals (WWF 2011). Fisheries were expected to take 10–20 years to recover, although not completely, from a cyanide spill from a gold mine into a river in Romania in 2000 (Batha 2000).

Indirect effects can result from the resources required to operate the mine. For example, processing of bauxite into alumina requires large quantities of electricity. Hydroelectric power plants will supply the first bauxite mine in Lao PDR (Ord 2006b, p. 9). Hydroelectricity is expected to be cheaper than the electricity used at competing mines elsewhere in the world. Hydroelectricity may also provide a carbon advantage in the future if carbon taxes and other controls are widely introduced to mitigate climate change. If absolute carbon reductions are mandated then not only will bauxite from Lao PDR, and perhaps Cambodia and Vietnam if production there can also access hydroelectricity, be cheaper but demand for low carbon alumina will strengthen the imperative to extract and process bauxite in this bauxite triangle of the wider Mekong region will intensify. Additional hydroelectric power plants may be needed. Hydroelectric power plants, as previously discussed, however threaten considerable damage to fisheries and agriculture, particularly when multiple dams present cascades of cumulative impacts across river basins, ecosystems and borders.

The challenge of mining for food security is that the industry is already developing ahead of the capacity of governments to legislate for the highest standards and enforce environmental management which would protect existing and future options for food production. Until regulation and enforcement catch up protection of food production will for much of the time until 2030 rest with the standards and commitments of the industry. It is a considerable challenge. The experience of mining in the Philippines left the Senate to conclude that without appropriate controls it is a grave threat to food security (Philippines Senate 2011).

However, if mining is well managed with all emissions and effluents properly treated and tailings secured then the revenues earned by the state may support the development of sustainable higher-yielding food production elsewhere which will outweigh the unavoidable aspects of mining's environmental and social footprint. Moreover, mining exports may support stronger currencies making food imports cheaper, although this benefit may be outweighed by the costs of this "Dutch disease" economic phenomenon.

Mining in Lao PDR is now well established as commercially viable. Bauxite follows the development of commercial mining of copper, gold which began a decade ago and more recently coal. Other minerals also abound (MPI 2008, p. 1); (Vientiane Times 2009) with 119 companies have interests in exploring or developing 193 mining sites (Barney 2007, p. 11). The potash, estimated at between 4.929 and 50 billion tonnes (DGM); (MPI 2008), of the Vientiane Plain is one of a handful of major deposits worldwide of the mineral used to make agricultural fertilizers. Given the resources, their proximity to major markets especially China, and a background of steady if not robust global demand more mines, bringing risks and rewards, are in prospect around Lao PDR, and perhaps Cambodia and Vietnam in the years to 2030. The risks for food security will remain high while governance remains underdeveloped.

Dimensions impact estimate 2030 – mining

Dimension	Impact
Availability	▪ Local production will be erased which may force farmers to marginal lands or dependence on wages and food markets ▪ Production elsewhere may suffer decline environmental quality or ecosystem services because of resources demanded by mines ▪ Pollution presents chronic and acute risks to local and regional production

(continued)

Dimensions impact estimate 2030 – mining	
Dimension	Impact
Accessibility	▪ Farmers may not be in control of land, resources or technologies used to mitigate effects of mining on agriculture ▪ Farmers will lose access to means of production due to land, water and resource demands of mines ▪ People whose livelihoods are damaged to ruined by mines will depend on labour markets for wages sufficient to afford strong food security from food markets
Utilization	▪ Crops, livestock and fish may be tainted with pollution such as heavy metals ▪ Ground and surface water used by agriculture and households may be polluted or insufficient
Stability	▪ Less local food production due to mining will leave communities more exposed to acute and chronic changes in regional and international food markets ▪ Water demands for generating hydroelectricity for mines may take precedence over demands of food production exacerbating effects of dry season or drought

Enabling factors impact estimate 2030 – mining	
Enabling factor	Impact
Demography	▪ Impact of mining labour force on local food demand may be significant ▪ Mining impact on local labour markets unclear ▪ Opportunities for farmers to switch into mining jobs may be limited ▪ Alternatively strong labour demand from mines might cause people to quit farming leaving fields idle
Resources	▪ Mining requires large areas of land, especially surface operations associated with bauxite ▪ Water demands for processing and energy generation may exceed tolerances of local watersheds ▪ Demands for other materials, such as aggregates and timber, may affect food production directly or indirectly elsewhere
Infrastructure	▪ Ecological – restructuring of the landscape, tailings, and water consumption for processing and energy generation risk major disturbance and degradation at local and regional scales ▪ Hard – energy, transportation, processing and tailings storage will alter the landscape affecting flow of ecological services and impeding access to fields and fisheries ▪ Hard – roads and railways may improve the flow and reduce cost of imported food ▪ Soft – knowledge and skills may be required for different farming livelihoods elsewhere or to take up jobs in mining
Markets	▪ Food, farming and farmers may have insufficient regulatory or pricing power to compete with mines in markets for resources ▪ Labour market regulation may be insufficient to ensure farmers and other people whose livelihoods are damaged by mining have access to high-wage jobs in mining

(continued)

Enabling factors impact estimate 2030 – mining	
Enabling factor	Impact
Policy	▪ Absence of markets for environmental goods and ecosystem services may obstruct sustainable management and use by mining and agriculture ▪ Large investments and industrial development emphasised ▪ Industrial promotion not harmonized with environmental and food policies ▪ Underdeveloped regulatory, management and enforcement institutions require priority attention of policy and resources to minimize environmental impacts and maximize public well-being including food security
Assessment	
Negative	▪ Inadequate governance ▪ Disappointing track record ▪ Acute and chronic risks ▪ Direct and indirect impacts ▪ Local and transnational effects ▪ Strong demand drivers ▪ Catalyst for hydropower and railway development ▪ Royalties and taxes might mitigate productions losses ▪ Strong regulation and environmental management could produce net positive impact

4 Cumulative Impact Assessment

The developments taking place or expected to take place in the wider Mekong region by 2030 will impact food security through effects on ecology, economy and society. The ecology provides the foundations for producing food. The economy is the device society uses to mobilise food production and distribute food through exchanges of goods and services. The efficiency, sustainability and resilience of food production and distribution for food security are vulnerable to changes wrought by developments which affect the quality or distribution of resources. The impacts, without mitigation or adaptation, of developments will accumulate effects on the ecology, economy and society which may not be linear. Furthermore, the accumulated effects will also compound or magnify the impacts of other developments. The size, scale and speed of such impacts are beyond the scope of this assessment. Instead the assessment considers the qualitative nature and interactions of the cumulative impacts.

The developments are taking place in a region with underdeveloped governance. Even if developments could be designed and operated without having undue impacts on food security this is probably unlikely because of weak governance. The multi-dimensional nature of food security cutting across fields of interests, responsibility and power in governments requires an unusual level of coordination. Even developed countries, such as the UK which depends on imports of food, have expressed concerns about the challenges of maintaining strong food security. Consequently, the impacts of developments on food security in the wider Mekong region are likely to remain a peripheral concern for several more years, possibly decades.

Impacts on food security are not always immediately visible because most of the effects are not produced by relatively infrequent acute events, such as a pollution spill or typhoon. The developments taking place in the wider Mekong region may take a decade or more to mature. Food production will not cease and markets will not be empty as the impacts of the developments accumulate over the next few decades. The effects of impacts, if monitoring is taking place, may not be discernible for even longer. Recent experience suggests that even if relative indicators of weak food security decline absolute numbers of people suffering weak food security in the wider Mekong region may not change much.

It is this slow-burn impact which causes difficulties in envisaging how a development for which a strong, albeit narrowly-defined, economic case can be made will in the future obstruct strong food security and contribute to reinforcing entrenched weak food security. Furthermore, accumulated impacts, particularly degrading and fundamentally changing ecosystems and the availability of soil and water, will eliminate some of the potential for sustainably increasing agricultural productivity. Such loss of option value is often not considered in assessments of the wider economic returns and impacts of a development.

Developments however are expected to promote growth in the economy of the wider Mekong region. Mining and mainstream dams, in particular, promise to make substantial contributions in royalties, taxes and other payments to national treasuries, especially in Lao PDR. However, increases in gross domestic product do not *ipso facto* produce strong food security. A fast-growing economy is not a rising tide which lifts all boats. A hot economy is often accompanied by faster inflation. Where wages fail to keep pace or outstrip food prices, especially as more people become dependent upon markets for food or to ensure certain nutritional or temporal aspects of strong food security, people may either be unable to escape weak food security or see the character of weak food security change or become vulnerable to weak food security.

Substantial increases in state budgets raise prospects for improving food security, particularly production and where necessary welfare payments. But whether prospects are realised depends on the interests, priorities and capabilities of the state for investing in sustainable, climate-resilient agriculture. Therefore, the cumulative impacts of developments are more likely, directly and indirectly, to be negative while governance is weak and unable to grasp the complexity of food security and the consequences of impacts accumulating years into the future.

Dimensions impact estimate 2030

Dimension	Impact
Availability	▪ Resources for food production will contract in quantity and quality in relative and absolute terms ▪ Biodiversity will decline because forests and other natural habitats are replaced by monocrop industrial plantations ▪ Water flow patterns will change producing rapid shifts in the environment which are likely to exceed the speed and capacity of adaptation of many species

(continued)

Dimensions impact estimate 2030

Dimension	Impact
	▪ Nutrient flows will decline while pollution flows, such as agro-chemicals and heavy metals, will increase especially in water ▪ Declines in food production as industrialization absorbs more land and water may be offset, at least partially, by new knowledge and technology to improve productivity ▪ Productivity improvements are unlikely to be sustainable if they depend on agro-chemical methods and hydrocarbons ▪ Threats to food production will coincide with the stress of demand from a rising population which is increasingly urban
Accessibility	▪ Some farmers and fishers will lose rights to land and rivers to developments ▪ Dependence upon imported agricultural inputs will rise and more farmers may lose rights and agency to corporations holding patents ▪ More farmers and fishers will depend on markets for more of their food security ▪ Relative reductions in people suffering weak food security may not be sustainable and absolute numbers could rise ▪ Labour markets and trends in wages and job security will increasingly determine accessibility
Utilization	▪ Water for food production, food preparation and household sanitation faces rising competition from industrial and urban demand for water ▪ Household water sources face increasing risks of pollution from agro-chemicals, heavy metals and mining wastes ▪ Salinity may be a problem for domestic supply in coastal areas and regions of large-scale intensive irrigation ▪ Changes in the composition of food may not improve food security if people do not understand its nutritional value and how to use it
Stability	▪ Current development trends risk serious damage and decline in ecosystem vitality and services critical to sustainability and resilience of food production in the wider Mekong region ▪ Impacts of current development trends coincide with climate change and increasing stress in global food production ▪ Vulnerability to acute shocks will increasingly be mediated by changes in household income relative to market prices for domestic and imported food especially in the cities ▪ Competing demands for water between households, food production and industry, particularly dams and plantations, could turn to conflict during droughts which may be more frequent and severe because of climate change ▪ States may not be able to avoid choosing between water for food and water for industry with cross-border consequences

(continued)

Enabling factors impact estimate 2030

Enabling factor	Impact
Demography	▪ Changes in land use also change livelihoods turning farmers into labourers or migrants to cities ▪ Demand for labour from industrial developments will be uneven and may offer little prospect for work adding much value which can justify higher wages to afford strong food security ▪ Fewer opportunities for smallholder farming may amplify loss of agricultural knowledge and skills as farmers and their children chase better opportunities elsewhere ▪ Rural communities may fray and crumble due to the impacts of changes in land use coinciding with the pressures of climate change and better opportunities for some in the cities
Resources	▪ Land, soil, water and rivers are being transferred from smallholder agriculture for food and material production to industrial uses serving regional urban markets and global commodities trade ▪ The Mekong delta food production environment and ecosystem faces great pressure and uncertainty from the coincidence of upstream demands for land and water for industrial projects and coastal impacts of climate change ▪ Diversion of resources from food to industry amid intensifying impacts of climate change and the demands of urbanizing societies brings the prospect of expanding vulnerability to weak food security ▪ Improvements in transportation may cut costs of importing bulk agricultural inputs, such as fertilizers, facilitating increases in productivity
Infrastructure	▪ Ecological – the flow of ecological services at various scales is being distorted by expanding industrial demand impairing food production and introducing risks of unforeseen impacts across ecosystems ▪ Hard – hydrological engineering for irrigation or defence against erosion and saline intrusion might increase food production in the short term but risks long-term consequences which will overwhelm initial returns ▪ Hard – operating and maintenance costs of hydrological engineering could compete for funds with more sustainable investments in resilient food production ▪ Hard – communications, energy and transportation infrastructure could support agriculture while damaging the ecosystem through demand for land and water as well as making access easier for loggers and poachers ▪ Soft – knowledge and skills are required for farmers and food systems to adapt to the great changes taking place in land and ecology because of industrial developments and climate change ▪ Soft – households losing access to land and water for food production or experiencing declines in quality or stability may need new knowledge and skills to take up jobs elsewhere which afford strong food security
Markets	▪ Enclosure and extraction of natural resources undermines the environment for food production and livelihoods which combined with urbanization will leave more people dependent upon markets for food ▪ Vulnerability to certain shocks will rise as society's food market dependency ratio increases without policy and regulation to develop food supply resilience

(continued)

Enabling factors impact estimate 2030	
Enabling factor	Impact
	▪ Shifts in environmental quality because of industrial development causing sharp declines in aquatic protein will push markets to find substitutes leading to additional environmental impacts as livestock production expands to supply protein ▪ Expansion of dependence upon markets for food will coincide with changes and expansion of labour markets, fuelled in part by an exodus of unskilled or low-skill labour from agriculture amid decline in environmental quality because of the current trajectory of low-quality industrial development ▪ Limited recognition or understanding of the option value of high environmental quality may not improve without development of markets for environmental quality including environmental payments
Policy	▪ Requirements for higher revenues to support development have captured policymaking which supports rapid large-scale enclosure or extraction of natural resources in excess of regulatory capacity ▪ Near term returns from natural resources development for revenue produces a feedback reinforcing support for privileging natural resources development over assessment of social costs or longer term environmental degradation ▪ The prospects for industrial development to produce impacts which cascade across systems or catalyse seemingly unrelated effects in other systems in the short or long term is not adequately considered in policy for promotion, approval or management ▪ The absence of comprehensive balanced assessment of the complexity of costs and benefits of options for developing natural resources is unlikely to foster sustainable outcomes particularly for providing strong food security for all citizens ▪ Contradictions are emerging downstream where one arm of the state facilitates resources development while another expresses concern over the challenges of mitigating and adapting to the consequences for food security ▪ Internal policy coordination appears weak and short of coherence to manage resources development in ways which are Pareto efficient for food security and national resilience over the long term ▪ Primacy of resources development and the peripheral place of food security in policy making are producing outcomes with transnational consequences which will challenge the pretence of amity in international relations within the wider Mekong region particularly between states facing the dynamics of upstream-downstream environmental power asymmetries ▪ Existing regional agencies with mandates for supporting sustainable development of the wider Mekong region lack the authority, resources and conceptual foundations to operate effectively and without fostering contradictions ▪ Environmental management is inadequate and falling behind the scale, speed, spatial effects, technology and interactions of industrial developments underway or planned in the wider Mekong region

(continued)

Development impacts on wider Mekong region food security	
Development	Impacts and reactions
Mainstream dams	▪ Disruption of flow patterns, quantity and quality of water ▪ Changes in silt loads and reduction in nutrients ▪ Elimination of fish migration passages ▪ Impacts accumulate downstream damaging Cambodian flood plain and Mekong delta ▪ Royalties and taxes paid to the state might support sustainable improvement and intensification of food production elsewhere ▪ Where adaptation is possible farmers and fishers will require more costly inputs which may further damage ecosystems and food quality ▪ Generates energy to facilitate mining and processing of rubber ▪ Stress on food production compounded by climate change, water diversion and rising seas
Water diversion	▪ Increases in food production unlikely to be sustainable ▪ Could exacerbate existing environmental problems especially salinization ▪ Efficiency and effectiveness unproven ▪ Insufficient labour and interest in farming livelihoods ▪ Long-term maintenance and management maybe unsustainable ▪ Conflict or compound impacts of dams on food production downstream
Rising seas	▪ Erosion, flooding and saline intrusion changing food production landscape ▪ Mitigation by engineering hard infrastructure is economically and ecologically unsustainable ▪ Hard infrastructure can exacerbate stress or change elsewhere in the food production landscape ▪ Interactions with cascading impacts from upstream developments unclear ▪ At risk from all upstream developments affecting water flow, quantity and quality, including agro-chemical and industrial pollutants
Rubber	▪ Erase or displace food production and foraging ▪ Disrupt and degrade ecosystem services which support food production ▪ Disturb and distort ground and surface water flow and quality which may leave less for food production and household use ▪ Compounds ecosystem impacts of dams and water diversion
Railways	▪ Limited direct or indirect impacts ▪ May support cheaper food distribution and supply security or facilitate extraction of food through trade ▪ Cut costs and improve profitability of other developments such as mining and plantations
Bauxite mining	▪ High demand for resources required for food production or which through their provision could affect food production ▪ Soil, water and air pollution can taint or destroy food production locally or regionally ▪ Comprehensive and powerful regulation required to prevent pollution and limit impacts of resources demand ▪ Royalties and taxes paid to the state might support sustainable improvement and intensification of food production elsewhere ▪ Could compound decline in water quality caused by dams, water diversion and rubber

(continued)

Resource impacts on wider Mekong region food production environment	
Resource	Impacts and reactions
Land (soil, fields, forests, grasslands)	▪ Mainstream dams – reservoirs ▪ Water diversion – infrastructure and salinization ▪ Rising seas – erosion, flooding, saline intrusion ▪ Rubber – plantation expansion ▪ Railways – limited ▪ Bauxite mining – access ore, processing, tailings storage
Water (ground, surface)	▪ Mainstream dams – retention, flow disruption, sediment trapping, nutrient losses ▪ Water diversion – exacerbate low flow in dry season ▪ Rising seas – contaminate groundwater, displace freshwater ▪ Rubber – reduce ground and surface water, agro-chemical run-off ▪ Railways – fuel and lubricant pollution ▪ Bauxite mining – processing and power generation, wastes and tailings
Ecosystem	▪ Mainstream dams – eliminate fish migration and reproduction, reduce nutrient flows to farms downstream ▪ Water diversion – disturb function by reducing water flow, biodiversity reduction due to salinization ▪ Rising seas – expand coastal and brackish water ecosystems ▪ Rubber – reduce biodiversity and undermine ecosystem resilience ▪ Railways – limited ▪ Bauxite mining – reduce biodiversity, disturb water flows and functions, diminish through pollution e.g. heavy metals

5 Conclusion

Food security in the wider Mekong region is unlikely to improve markedly, that is in terms of absolute numbers suffering the deleterious effects of weak food security, in the years to 2030. The developments taking place at large scale are fundamentally changing the rich landscape and diverse ecology which underpins food production and which holds the potential to sustain and perhaps even increase food production despite climate change. The impacts of some developments, such as mainstream dams and indeed any dams on one of the world's most diverse river ecosystems and certainly its most productive, cannot be mitigated nor adapted to in ways which will come anywhere close to compensating for losses in food output and household food security, livelihood and community resilience. Others such as bauxite mining and rubber production might with research and comprehensive regulation proceed in ways which do not degrade ecosystems and food production or human well-being in general.

To realise the potential benefits of developments without compromising food security requires a degree of governance which is currently absent and unlikely to be in place throughout the wider Mekong region by 2030. Moreover governance is shaped by prevailing ideas and paradigms of what constitutes appropriate development in the eyes of power holders, primarily states. Currently, development is enthused with a sense that the environment is a set of building blocks which may be rearranged to yield a desired benefit with little consequence.

However, the prevailing reductionist perspective misses the benefits which flow from the arrangement of building blocks which interact in a system. Strong food security requires recognition of where food comes from, its origin as a product of ecosystems. Strong food security may be possible when the wider Mekong region maintains and nurtures the ecosystem to sustain and increase its economic value, while pursuing developments, such as rubber and mines, without compromising the vitality and resilience of the ecosystem.

Prospects for food security in the wider Mekong region will improve if large dams are abandoned and dismantled, water diversion plans are scrapped, adaptation meets rising seas, rubber and pulp are produced by polycropping, and mining for bauxite or anything else is tightly regulated to maintain environmental quality.

References

Anderson, K., and A. Bows. 2011. Beyond 'dangerous' climate change: Emission scenarios for a new world. *Philosophical Transactions of the Royal Society A* 369: 20–44.

Baird, I. 2010. Land, rubber and people: Rapid agrarian changes and responses in Southern Laos. *Journal of Lao Studies* 1(1): 1–47.

Baird, I., and B. Shoemaker. 2007. Unsettling experiences: Internal resettlement and international aid agencies in Laos. *Development and Change* 38(5): 865–888.

Baran, E. 2006. *Fish migration triggers in the lower Mekong basin and other tropical.* Vientiane: Mekong River Commission.

Barney, K. 2007. *Power, progress and impoverishment: Plantations, hydropower, ecological change and community transition in Hinboun district Lao PDR.* Toronto: York Centre for Asian Research.

Batha, E. 2000. Death of a river. *BBC*, February 15.

Baumann, H., S. Talmage, and C. Gobler. 2011. Reduced early life growth and survival in a fish in direct response to increased carbon dioxide. *Nature Climate Change* 2: 38–41.

BBC. 2010. Hungary battles to stem torrent of toxic sludge. *BBC*, October 5.

Bruijnzeel, L. 2004. Hydrological functions of tropical forests: Not seeing the soil for the trees? *Agriculture, Ecosystems and Environment* 104: 185–228.

Calder, I. 2003. *Tropics afforestation water infiltration perceptions.* Newcastle: University of Newcastle-Upon-Tyne.

Chungyalpa, D. 2010. Testimony of Dekila Chungyalpa Director, Greater Mekong program, World Wildlife Fund on challenges to water and security in Southeast Asia for the Senate Committee on Foreign Relations Subcommittee on East Asia and Pacific. United States Senate, Washington, DC.

Collins, N. 2011. Sea levels rising at fastest rate in 2,000 years. *The Daily Telegraph*, June 21.

Department of Forestry. 2007. *Forestry sector development report for 2006/2007 (draft).* Vientiane: Ministry of Agriculture and Forestry.

DGM. (n.d.). Mineral potential in Lao PDR. Retrieved 5 Mar 2012, from Department of Geology and Mines: http://www.dgm.gov.la/.

Doyle, A. 2011. Seas could rise up to 1.6 meters by 2100: Study. *Reuters*, May 3.

Eastham, J., F. Mpelasoka, M. Mainuddin, C. Ticehurst, P. Dyce, G. Hodgson, et al. 2008. *Mekong river basin water resources assessment: Impacts of climate change*. Canberra: Commonwealth Scientific and Industrial Research Organisation.

FAO. 2006. *Policy brief issue 2 – Food security*. Rome: Food and Agriculture Organzation.

Foppes, J., and S. Ketphanh. 2007. *NTFP use and household food security in Lao PDR*. Vientiane: Forest Research Centre.

Frommel, A., R. Maneja, D. Lowe, A. Malzahn, A. Geffen, A. Folkvord, et al. 2011. Severe tissue damage in Atlantic cod larvae under increasing ocean acidification. *Nature Climate Change*. doi:10.1038/nclimate1324.

Fujita, Y. 2007. *Key issues in smallholder rubber planting in Oudomxay and Luang Prabang provinces, Lao PDR*. Vientiane: National Agriculture and Forestry Research Institute.

Fullbrook, D. 2010. Food as security. *Food Security* 2(1): 5–20.

Fullbrook, D. 2011. *Smallholder production agreements in the Lao PDR: Qualifying success*. Vientiane: Laos Extension for Agriculture Project/Helvetas.

Fullbrook, D. 2012. Averting a dam tragedy: The Mekong as a model for common asymmetric resources management.

Galland, M. 2007. *Consuming the future – The real status of biodiversity in Lao PDR*. Vientiane: IUCN, WCS and WWF.

Gao, J. 2005. Australian mining company denies poisoning villagers. *Sydney Morning Herald*, August 10.

Gregory, J., and J. Lowe. 2010. A sea of uncertainty – How well can we predict future sea level rise? *Nature Reports Climate Change* 4: 42–43.

Hansen, J., and M. Sato. 2011. *Paleoclimate implications for human-made climate change*. New York: NASA Goddard Institute for Space Studies/Columbia University Earth Institute.

Hanssen, C. 2007. *Lao land concessions, development for the people?* Vientiane: CIDSE-Laos.

Hori, H. 2000. *The Mekong: Environment and development*. Shibuya: United Nations University Press.

ICEM. 2010. *Mekong river commission strategic environmental assessment of hydropower on the Mekong mainstream – final report*. Hanoi: International Center for Environmental Management.

IPCC. 2007. *IPCC* Fourth assessment report: Climate change 2007: Working group I: The physical science basis. Retrieved 26 Feb 2012, from intergovernmental panel on climate change: http://www.ipcc.ch/publications_and_data/ar4/wg1/en/spmsspm-projections-of.html.

Johnston, R.M., C.T. Hoanh, G. Lacombe, A.N. Noble, V. Smakhtin, D. Suhardiman, et al. 2010. *Rethinking agriculture in the greater Mekong subregion: How to sustainably meet food needs*. Columbo: International Water Management Institute.

Käkönen, M. 2008. Mekong delta at the crossroads: More control or adaptation? *Ambio* 37(3): 205–212.

Keeratipipatpong, W. 2011. Thai food security under threat. *Bangkok Post*, May 12.

Le, Q.M. 2001. Environmental governance: A Mekong delta case study with downstream perspectives. In *Mekong regional environmental governance: Perspectives on opportunities and challenges*, ed. N. Badenoch, and M. Dupar. Resources Policy Support Initiative. World Resource Institute, Washington DC.

Le, T.V., H.N. Nguyen, E. Wolanski, T.C. Tran, and S. Haruyama. 2007. The combined impact on the flooding in Vietnam's Mekong river delta of local man-made structures, sea level rise, and dams upstream in the river catchment. *Estuarine, Coastal and Shelf Science* 71: 110–116.

Lowe, J., T. Howard, A. Pardeans, J. Tinker, G. Jenkins, J. Ridley, et al. 2009. *UK climate projections science report: Marine and coastal projections*. London: Department for Environment Food and Rural Affairs.

Mineral Policy Institute. 2005. *Phu Bia mine – already in trouble*. Perth: Mineral Policy Institute.

Molle, F. 2005. *Irrigation and water policies in the Mekong region: Current discourses and practices*. Colombo: International Water Management Institute.

Molle, F., and P. Floch. 2008. Megaprojects and social and environmental changes: The case of the Thai "water grid". *Ambio* 37(3): 199–204.

MPI. 2008. *2008–2020 Planning for industrial economic development and cooperation in northern part of Lao people's democratic republic (Mid-term draft for opinions).* Vientiane: Ministry of Planning and Investment.

Nesbitt, H., R. Johnston, and M. Solieng. 2003. Mekong river water: Will river flows meet future agriculture needs in the lower Mekong basin? In *Water in agriculture – proceedings of a CARDI international conference on research on water in agricultural production in Asia for the 21st century*, ed. V. Seng, E. Craswell, S. Fukai, and K. Fischer, 86–104. Phnom Penh: Australian Centre for International Agricultural Research.

New Scientist. 2008. Sea levels 'will rise 1.5 metres by 2100'. *New Scientist*, April 16.

Nguyen, V.N. 2010. Food security and economic development in Vietnam. *International conference on revisiting agrarian transformation in Southeast Asia.* Chiang Mai.

Nhan, D.K., V.A. Phap, T.H. Phuc, and N.H. Trung. 2012. *Rice production response and technological measures to adapt to salinity intrusion in the coastal Mekong delta.* Can Tho: Can Tho University.

Niels Bohr Institute. 2011. *Sea levels will continue to rise for 500 years.* Retrieved 26 Feb 2012, from Niels Bohr Institute: http://www.nbi.ku.dk/english/news/news11/sea_levels_will_continue_to_rise_for_500_years_/, October 18.

Norway. 2006. *Two companies – Wal-mart and freeport – are being excluded from the Norwegian government pension fund – Global's investment universe.* Retrieved 6 Mar 2012, from Ministry ofFinance:http://www.regjeringen.no/en/dep/fin/Press-Center/Press-releases/2006/Two-companies---Wal-Mart-and-Freeport---.html, June 6.

Ord. 2006a. *Developing an integrated aluminium industry in Laos.* Sydney: Ord River Resources.

Ord. 2006b. *Laos Bolaven plateau bauxite project update.* Sydney: Ord River Resources.

Pegg, S. 2006. Mining and poverty reduction: Transforming rhetoric into reality. *Journal of Cleaner Production* 14(3–4): 376–387.

Philippines Senate. 2011. *Irresponsible mining threatens food security.* Retrieved 5 Mar 2012, from Senate of the Philippines: http://www.senate.gov.ph/press_release/2011/1208_pangilinan2.asp, December 8.

Qiu, J. 2009. Where the rubber meets the garden. *Nature* 457(7227): 246–247.

Redsludge. 2010. *Current status report.* Retrieved 5 Mar 2012, from Redsludge: http://redsludge.bm.hu/?p=446, October 20.

Rigg, J. 2006. Forests, marketization, livelihoods and the poor in the Lao PDR. *Land degradation and development* 17(2): 123–133.

Rignot, E., I. Velicogna, M.R. van den Broeke, A. Monaghan, and J. Lenaerts. 2011. Acceleration of the contribution of the Greenland and Antarctic ice sheets to sea level rise. *Geophysical Research Letters* 38(5): L05503.

Rowe, M. 2011. Come hell or high water. *Geographical* 83(4): 34.

Shi, W. 2008. *Rubber boom in Luang Namtha – A transnational perspective.* Vientiane: Deutsche Gesellschaft für Technische Zusammenarbeit.

Simutanyi, N. 2008. *Copper mining in Zambia – The developmental legacy of privatization.* Pretoria: Institute of Security Studies.

Sithong, T., and Thoumthone, V. 2006. Impacts of cash crops on rural livelihoods: A case study from Muang Sing, Luang Namtha province, northern Lao PDR. In RECOFTC, *Hanging in the balance: Equity in community-based natural resource management in Asia.* Honolulu: East–west Centre.

Sumaila, U.R., W.W. Cheung, V.W. Lam, D. Pauly, and S. Herrick. 2011. Climate change impacts on the biophysics and economics of world fisheries. *Nature Climate Change* 1: 449–456.

SWGUD. 2008. *Policy brief #2: Rural land management and land administration – DRAFT.* Vientiane: Ministry of agriculture and forestry, sub-working group on uplands development.

Thukral, N. 2011. Top 10 rice exporting, importing countries. *Reuters,* January 28.

Toan, T.Q., N.H. Trung, and D.K. Nhan. 2011. *The Mekong future project: first draft on the hydrological simulation.* draft, Can Tho: Can Tho University.

VFA. 2012. *Vietnam Jan 01 – Dec 31, 2011 Rice exports 7.105 MMT.* Retrieved 12 Jan 2012, from Vietnam Food Association: http://www.vietfood.org.vn/en/default.aspx?n=5911, January 9.

Vientiane Times. 2008. Ministry protects forests from rubber plantations. *Vientiane Times,* November 18.

Vientiane Times. 2009. Large gold reserves augur a bright future. *Vientiane Times,* May 29.

WFP. 2007. *Lao PDR: Comprehensive food security and vulnerability analysis*. Rome: World Food Programme.

Winwong, K. 2008. River diversion plans for whose benefit? *Inter-Press Service,* August 4.

World Commission on Dams. 2000. *Dams and development: A new framework for decision-making*. London: World Commission on Dams/Earthscan.

WSFS. 2009. *Declaration of the world summit on food security*. Rome: World Summit on Food Security.

WWF. 2011. *One year after the red mud disaster: WWF warns against ticking time bombs, calls for action plan for Central and Eastern Europe.* Retrieved 5 Mar 2012, from WWF: http://www.wwf.eu/?201848/One-year-after-the-red-mud-disaster-WWF-warns-against-ticking-time-bombs-calls-for-action-plan-for-Central-and-Eastern-Europe, October 3.

Ziv, G., E. Baran, S. Nam, I. Rodriguez-Iturbe, and S. Levin. 2012. Trading-off fish biodiversity, food security, and hydropower in the Mekong river basin. Proceedings of the National academy of sciences of the United States of America.

Chapter 4
Impacts of Natural Resource-Led Development on the Mekong Energy System

Tira Foran

1 Introduction

How might mainstream hydropower dams, irrigation, sea-level rise, upland rubber, enhanced rail connectivity, and smelting aluminium affect the energy system of mainland Southeast Asia? Upon considering impacts, what implications emerge for sustainable natural resource and infrastructural development? This chapter describes the Mekong region's energy system, focusing on how six sets of potential changes introduced in Chap. 1 may impact on it. We approach the energy system through eight dimensions: energy metabolism, governance, availability, access, affordability, long-term sustainability of supply, resilience, and environmental sustainability. The analysis identifies possible impacts of each of the six potential developments on the system, individually as well as cumulatively.

The analysis includes original quantitative impact estimates, based on in-depth literature review, supplemented by the author's participation in policy debates in the Thai power sector, 2006–10 (Foran forthcoming). The chapter begins by situating the energy sector in geographic context. Section 2 describes the scale and dynamics of the energy system in more detail. Section 3 reviews key institutional and governance drivers of Mekong energy, while Sect. 4 outlines methods. Sections 5 and 6 present the impact assessment.

As noted in Chap. 1, the 'Mekong region' as defined in this book is substantially equivalent to the ADB's 'Greater Mekong Subregion' (GMS), except that it does not include the Guangxi Zhuang Autonomous Region of China. Table 4.1 compares Guangxi to other countries and regions of the GMS. In 2009 Guangxi had approximately 15% of the GMS population and 19% of its GDP. The Mekong region is projected to grow rapidly. For Thailand and Vietnam, the two largest economies of the region, an APEC data source projects GDP as growing 4.5% and 6.2% p.a. respectively during the period 2005–2030. Primary energy demand is projected as growing 3.0% and 3.8% respectively during the same time period (Asia Pacific Energy Research Center 2009).

A. Smajgl and J. Ward, *The Water-Food-Energy Nexus in the Mekong Region: Assessing Development Strategies Considering Cross-Sectoral and Transboundary Impacts*,
DOI 10.1007/978-1-4614-6120-3_4,

Table 4.1 Greater Mekong subregion – selected indicators 2009

	Cambodia	Lao PDR	Myanmar	Thailand	Vietnam	Yunnan	Guangxi	GMS
Population 2009 (million)[1]	14.0	6.1	47.7	68.7	87.0	45.9	46.0	315.4
GDP 2009 (billion USD 2000)[2]	7.4	3.2	19.9	173.9	58.8	56.0	74.9	394.1
Share of regional GDP	2%	1%	5%	44%	15%	14%	19%	100%
GDP/capita (USD 2000)	532	518	417	2,531	676	1,219	1,630	1,250
GNI/capita (PPP, current interna-tional dollars)	1,980	2,330	1,770	7,610	2,870	6,870 (China)		
Total primary energy supply (TPES) 2009 (PJ)[3]	218	211	633	4,339	2,484	2,106	2,340	12,331
TPES/Capita (GJ)	15.54	34.39	13.27	63.15	28.54	45.89	50.92	39.09
Net import depend-ency ratio	30%	15%	−48%	46%	−18%	5%	12%	42%
Electricity generated (TWh)	1.206	tbd	5.85	148.389	83.191	69	84	425

Source: Author, based on 1. UN (2012); 2. World Bank (2012b); 3. IEA (2011)
Note: TPES and Electricity values for Laos PDR, Yunnan and Guangxi estimated from IRM (2008: Appendices)

The region is part of ASEAN. International Energy Agency's reference scenario projected ASEAN's primary energy demand as growing 2.5% per annum between 2007 and 2030, significantly higher than the average rate in the rest of the world. In this scenario, ASEAN demand for electricity grows even more rapidly, 4.2%, during the same period (IEA 2009).

Ranked in terms of GDP (purchasing power parity) per capita, the Greater Mekong region's most materially prosperous economy in 2009 was Thailand, followed by China, Vietnam, Laos, Cambodia, and Myanmar (Table 4.1). Such indicators however mask important sub-national and cross-class disparities in income and well-being. The Table shows that the region is a net energy importer, although notable exporters include Vietnam and Myanmar. Vietnam's status as a net oil exporter however is in large part as a result of its limited refining capacity. Rapid growth in demand means that it will become a net oil importer by 2015.[1]

What role the energy sector plays in broader economic development – particularly rural development – remains a contested and complex topic. Mainstream analysts point to the positive relationship between energy use per capita and the Human Development Index. The relationship takes the form of a 'saturation' function, with the highest change in HDI occurring at low levels of energy consumption (United Nations 2010). This relationship is frequently used to support prescriptions to convert rural people away from using 'traditional' biomass. Delivering modern energy to rural people often has been taken to mean substitution of fossil fuels and the extension of national electricity networks (Greacen and Greacen 2004). But it could just as well mean village-scale systems powered by locally available water, solar, or biomass resources.

2 Scale and Dynamics of Energy System

Whatever position is taken regarding appropriate solutions for rural energy, we also need to appreciate that the Mekong energy system is very large, highly connected to global energy systems, as well as internally differentiated. A substantial overlap between our definition of 'region' and the GMS allows us to refer to detailed analysis commissioned by the ADB as part of its vision for regional integration (Asian Development Bank 2009). Integrietes Ressourcen Management (henceafter 'IRM') an Austrian consultancy, produced a large cost optimization model including energy for stationary and transport systems (2008). IRM explored scenarios involving incremental changes to policy with respect to regional cooperation as well as a scenario in which users are charged for certain environmental and social costs.

The IRM (2008) study is an important, accessible data source and compilation of energy scenarios. However the IRM model has a number of limitations. GDP growth (set exogenously, and derived from official estimates) drives useful energy demand via the assumption that useful energy intensity (energy consumed per unit of GDP) improves at a rate of 1.3% per annum, which is a better rate than global long-term average of 1% (IRM 2008: 56, 250). This derivation of useful energy demand is

[1] ASEAN's only other net oil exporters are Brunei and Malaysia. By 2030, only Brunei remains a net exporter. ASEAN oil production has been declining since 2000. By the year 2030, ASEAN oil imports as a fraction of total consumption may reach 74% (IEA 2009: 550).

more convenient, but much less informative than a bottom-up derivation based on trends in the end-use sectors. Second, the study lacks comprehensive disaggregated analysis of end-use energy efficiency options.

Third, despite being a cost optimization study, producer or consumer unit energy costs are not reported in a manner amenable to deriving affordability indicators. The analysis does not explore alternative scenarios around social transition. Importantly for the GMS, there is lack of clear discussion around strategic management of biomass resources, including priority setting between competing uses of biomass such as electricity vs. transport fuels vs. household cooking and other heating demands (Figure 4.2; Kahrl et al. 2012).

The above limitations mean that specific results from IRM (2008) presented below need to be interpreted with caution. For example, if the GMS region's GDP growth rate deviates significantly from 6.3% per annum – a rate assumed by these analysts before the global financial crisis – overall levels of energy demand will be different.[2] However the mix of energy technologies and activities presented is *indicative* of how the region might change, absent any marked transition in areas such as policy support for end-use energy efficiency and renewable energy.

2.1 Trajectories

Coal, oil, and gas accounted for ¾ of total primary energy supply (TPES) between 2001 and 2005. This chapter uses the IRM 'GMS Integrated' simulation as a reference scenario and figures in this section are based on those simulation. (Specific assumptions are discussed in Sect. 4 below).

Figure 4.1 projects primary energy supply 2005–25 (IRM 2008: Table 142). Average annual growth rate of TPES during this period is 4.9%, almost double the IEA projections for ASEAN 2007–30 (IEA 2009). By 2025 annual growth in TPES is estimated to moderate to 4.4%.

Figure 4.1 shows the continuing importance of biomass in contribution to total primary energy supply. Between 2005 and 25 biomass supply grows in absolute terms (by 48%) but declines in relative importance. The top three sources of primary energy in 2005 were crude oil, biomass, and coal; by 2025 they are estimated to be coal, crude oil, and biomass. Although the GMS biomass resource potential is estimated at 113,238 PJ/annum, more than four times the total primary energy demand in 2025, the biomass used in 2025 is less than four percent of this potential (IRM 2008: Tables 13, 142). The share of hydroelectricity in the region's primary

[2] This is the modelers' 'Low' case defined before the global financial crisis, equating to a 6% p.a. increase in useful energy demand (IRM 2008:53).

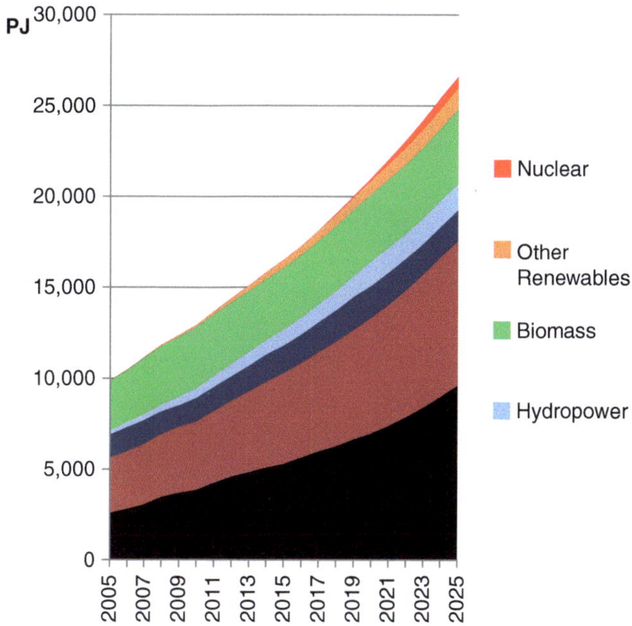

Fig. 4.1 Greater Mekong estimated primary energy demand, 2005–2025, (10^{15} J) (Source: Author, based on IRM (2008, Table 142). Primary energy of hydropower estimated based on potential energy of water held behind dams (equivalence method))

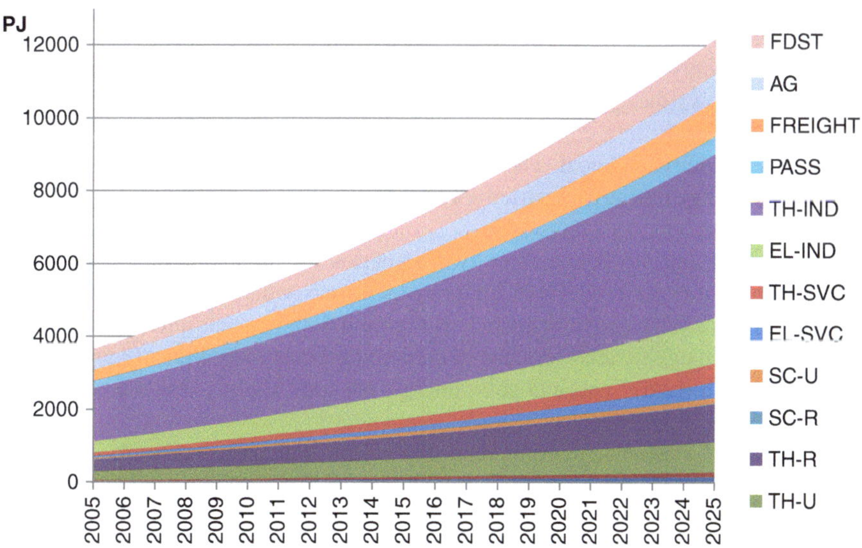

Fig. 4.2 Useful energy demand, 2005–2025, GMS Integrated scenario (10^{15} J) (Source: Author, based on IRM (2008: Table 137). *EL* electricity, *R* rural households, *U* urban households, *TH* thermal energy, *SC* space conditioning, *SVC* services, *IND* industrial, *PASS* passenger transport, *FDST* feedstocks)

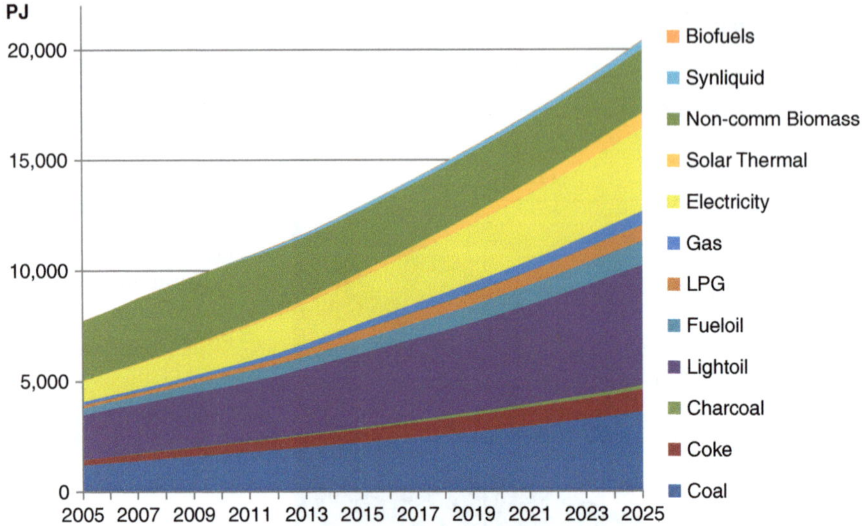

Fig. 4.3 Final energy demand, 2005–2025, GMS Integrated scenario (10^{15} J) (Source: Author, based on IRM (2008: Table 140). Note: Non-comm = non-commercial)

energy supply is estimated to grow from 2% to 5.5% during 2005–25, an annual growth rate of almost 10% (IRM 2008: Table 142).

Figures 4.2, 4.3 and 4.4 show how downstream components of the Mekong energy system might evolve. Industrial heating demand is the largest category of useful energy demand (Fig. 4.2). Industrial energy demand significantly exceeds household energy demand, which in turn exceeds transport energy demand. Figure 4.3 shows the growth of final energy demand. In 2005 the top three final energy carriers were non-commercial biomass, light oil, and coal. In 2025 they are light oil, electricity, and coal. The change in importance reflects modernization of household energy carriers (electricity, LPG) along with unabated demand for light oil for transport.

Figures 4.4 and 4.5 below show two scenarios for the evolution of electricity generation. Figure 4.4 is the GMS Integrated scenario, where certain environmental and social costs have not been included. Figure 4.5 shows that internalizing those costs results in significantly less use of coal and biomass, significantly greater use of hydropower, natural gas, and solar energy, but only slightly less use of nuclear power.

Pollutants for which environmental and social damage costs were estimated included carbon dioxide, methane, nitrous oxide, nitrogen oxides, and sulfur dioxides, as well as $/MWh estimates of damages from hydropower (IRM 2008: 50–51). The 'Low Carbon' scenario phased in the damage costs incrementally, adding 10% of the full estimated cost per year over a 10-year period beginning 2005. The average price of energy after internalizing the damage costs was approximately 15% greater than not internalizing (IRM 2008: 228). This scenario also assumed a –0.2%

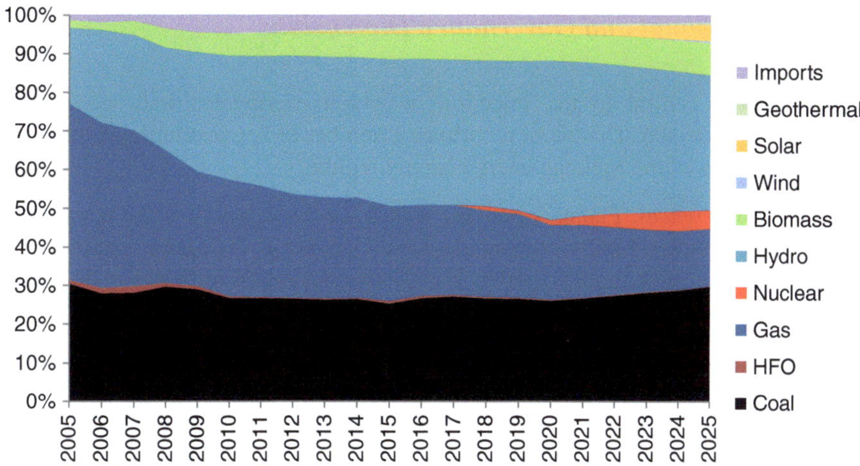

Fig. 4.4 Power generation mix, 2005–2025, GMS Integrated scenario (Source: Author, based on IRM (2008: Table 141))

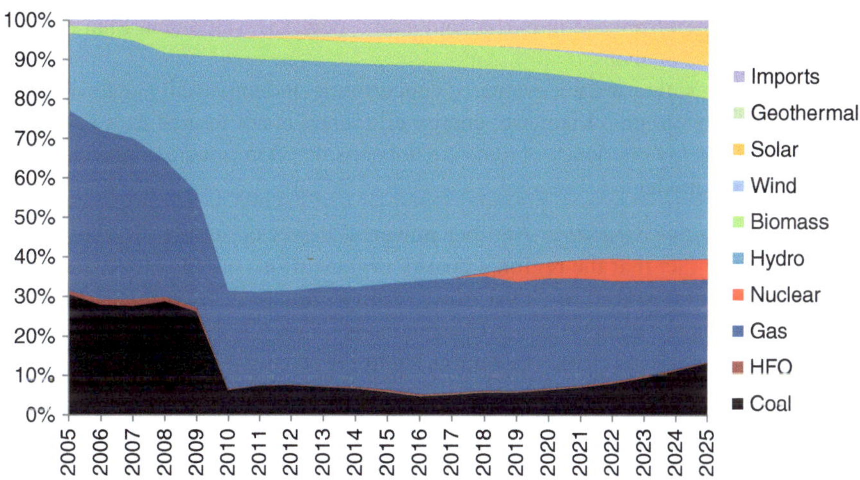

Fig. 4.5 Power generation mix, 2005–2025, GMS Low Carbon scenario (Source: Author, based on IRM (2008: Table 141))

price elasticity of demand. The effect of including certain environmental and social costs is to reduce total demand slightly (by <5%). The direct effect of including environmental and social costs on the entire energy portfolio is to add only $13 billion dollars to the discounted total portfolio cost of $1,072 billion (IRM 2008: 227). This relatively small difference reflects the specific assumptions about the magnitude of the damages and their phasing in.

2.2 *Institutional and Governance Dimensions*

Governance is central to the selection of which energy/electricity generation pathways are pursued. This section outlines a number of key institutional and governance features of the Mekong energy supply regime.

Technical Preferences. Planners perceive that large-scale supply is the most cost-effective way to meet rapidly growing demand (Foran et al. 2008). For instance, the computer optimization models used by Electricity Generating Authority of Thailand include the following options to meet rising demand (EGAT 2007):

700 MW coal-fired thermal power plant;
700 MW gas-fired combined cycle power plant;
230 MW gas turbine power plant; and
1,000 MW nuclear power plant

Until recent diversification decisions, the Thai power utility compared imported hydropower solely on a cost basis against the cost of domestic combined cycle natural gas turbine.

Smaller-scale options, such as <100 MW biomass combined heat and power (CHP) plants are introduced into Thailand's final Power Development Plan ex post cost-optimization analysis.

In other words, they are not evaluated concurrently and impartially against large-scale generation options. Likewise, energy efficiency is not treated as a resource, meaning that energy efficiency options are not considered in candidate power plants in the planning model.

Financial Incentive Structure. Another powerful driver of supply-side solutions stems from the fact that the region's energy organizations operate according to a traditional rate-of-return or "cost-plus" utility model. Under this structure, profitability hinges directly on gross revenues. There is limited incentive to invest in energy savings, and maximum incentive to invest in generation plants owned by the utility.

Meaningful Participation. Decision-making by government elites, with limited or negligible public input into a range of important issues, is common in the region. Such issues range from locating power station sites to major privatization initiatives (Greacen and Greacen 2004; Foran 2006; Nakhooda et al. 2007).[3] Thai governments have had significant and recurrent problems with gaining public acceptance of new power stations and gas pipelines. Coal is particularly controversial in Thailand, but even small-scale biomass plants have proven difficult. These difficulties are driven by the lack of impartial local consultation processes, occasional corruption of local government, and the perception by civil society organizations that environmental impact assessment processes have little bearing on decision making.

[3] In the United States, participation in the affairs of energy organizations occurred as a response to crisis in public utility financing, in the wake of the first and second oil crises.

In this context, a review of Mekong hydropower governance found four key sets of development and governance issues (Foran et al. 2010), which together define the "Mekong hydropower regime:" (1) resilience of the Mekong's aquatic ecosystems, in particular its fisheries, which are uncertain but expected to decline as more dams are built throughout the region; (2) structure of the electricity industry, in particular an electricity supply chain dominated by monopolistic state utilities, in which willingness to pay, consumer choice, and awareness of sustainability issues are limited; (3) hydropower host country regulation, which has limited accountability to citizens and is constrained in terms of technical and legal enforcement capacity; and (4) "bottom-up" river basin development, where a range of project sponsors or proponents (multilateral and private financiers and developers), confront relatively weak state regulatory practices. Among various project sponsors, we can detect higher- and lower-risk project design and investment decisions with respect to environmental and social impact mitigation.

3 Method

3.1 Approach to Energy Security

A number of indicators selected from literature on energy security support our analysis. Table 4.2 below presents 13 indicators, representing eight important dimensions of energy security. The selection reflects a compromise between comprehensiveness and maintaining a coherent focus on impacts of development on energy. Because the energy system is embedded in wider social-ecological systems, energy security is contested. The concept has been dominated by concerns with security of supply (short-term availability) and access at affordable prices. In more recent years analysts have called for adding environmental and socio-political attributes to the concept (Vivoda 2010; Sovacool et al. 2011).

For example, for environmental sustainability, in addition to metrics such as CO_2 and SO_2 emissions per capita, Sovacool et al. (2011) also include national forest cover and freshwater availability. However, including such metrics raises questions about causality and coherence. Non-energy developments also impact on forest cover and water availability (e.g. irrigation schemes may increase local water availability; rubber plantations and upland intensification may reduce forest cover). Although important, such broader system indicators are deliberately not included in our energy impact assessment. The reader interested in additional social-ecological indicators should refer to the relevant assessment chapter.

The diversity and complexity of the Mekong energy system challenges indicator-based assessment. In data-poor situations, the use of single metrics is pragmatic but also dangerously reductive. Poorly understood dimensions include how access to and affordability of different energy sources will change over time, particularly for rural people . Here it was necessary to make a number of assumptions, as we discuss in Sect. 2 below.

3.2 Reference Scenario

A Reference energy scenario provides a baseline and trajectory against which to assess incremental and cumulative impacts. As with all scenarios, it is meant to be read as an important and plausible set of factors and associated trends. For rates of economic growth, energy demand compositions and trajectories, our Reference scenario draws on the IRM 'GMS Integrated' scenario. Several assumptions and modelling choices influence the GMS Integrated scenario and hence shape the Reference scenario. First, impacts of climate change are not explicitly accounted for in the GMS Integrated scenario. For the purpose of this assessment some relevant climate impacts are summarized in Table 4.3. In Sect. 5 below these impacts are used to contextualize irrigation and rubber expansion impacts on the energy system.

Second, the GMS Integrated scenario assumes liberalized energy trade between all members of the GMS, such that economic cost of energy supply equals to production and delivery cost of energy, not equal to world market prices (IRM 2008: 72). Third, environmental and social damage costs are either not factored into the cost optimization modelling, or (in the case of the Low Carbon scenario) done in way difficult to interpret. Defining a Reference scenario based on the above policy settings should not be interpreted as a policy endorsement. The prospect of increased regional energy trade has been particularly controversial among observers concerned about its ecological and social implications (Greacen and Palettu 2007; Sovacool 2009). Increased energy integration in the Reference scenario is nonetheless consistent with the contemporary aspirations of the Mekong region countries.

Further assumptions were necessary to structure the Reference scenario: for *Contention* (Indicator 3), we assumed roughly that region-wide, one in five large coal, hydropower, natural gas, or nuclear power stations is the object of dispute lasting more than 1 year (i.e. a mobilization of claims on behalf of aggrieved people against authorities). This is based on the observation that the comparable fraction in Thailand has been greater than one in three.

We chose *Hydropower supply* (Indicator 4) as an indicator of Availability. Additional indicators such as regional *sufficiency* in primary energy (i.e. regional production/total supply) would be more comprehensive, but unfortunately the GMS Integrated scenario did not report imports and exports to 2025 in sufficient detail. However we assess *Years of remaining fossil energy reserves* (Indicator 5). This indicator is computed as the region's remaining Grade 1 (lowest-cost) coal, oil, and gas reserves, divided by primary energy consumption in a given year, with regional production shares taken from IRM (Table 4.4 below).

For *Access to electricity in low access Mekong countries* (Indicator 7), we provide two metrics showing average access levels in Cambodia and Myanmar. These levels are taken from the World Bank's "Business-as-Usual" (2011) scenario for those two countries, reflecting, in the case of Cambodia, slow progress mobilizing needed capacities, despite high rates of economic growth. For *Proportion of population in Mekong countries unable to afford modern energy services* (Indicator 8), we estimated the poverty headcount assuming current income distributions.

Table 4.2 Impacts of climate change on regional temperature and Mekong basin water

Variable	Infrastructure development case	Expectation with climate change, including specified infrastructure
Temperature		~ 0.023 °C/year temperature increase
Changes to Mekong flow at 14 key stations, 2010–2050	Baseline (year 2000 infrastructure)	Range of 4–9% increase in mean annual flow[a]
		Range of 10.6–29.5% increase in mean low-season flow
		Range +270 m^3/s (Chiang Saen) to +810 m^3/s (Stung Treng), low-season flow
	MRC 20-year foreseeable future scenario[b]	Range 4.2–14% increase in mean annual flow
	804,000 ha increase in irrigated area basin-wide (of which 316,000 ha in Thailand)	Range 7.5–16.4% increase in mean low-season flow
		Range +160 m^3/s (Chiang Saen) to +583 m^3/s (Kratie), low-season flow

Source: Hoanh et al. (2010: 53–54)
[a]Range across two IPCC climate scenarios, A2 (more extreme) and B2
[b]Includes hydropower (Chinese dams, mainstream dams, and tributary dams), irrigated area expansion, and domestic and industrial water demands (Hoanh et al. 2010: Table 2.5)

Table 4.3 Grade 1 non-renewable energy resources and regional production share of total consumption

	PJ	Cost (USD/GJ)	Regional production share of consumption (2001–2025) (%)
Coal	243,513	1.6	77
Lignite	310,271	1.5	77
Crude oil	18,900	0.8–2.2	57
Natural gas	9,310	1.7	100

Source: IRM (2008: 10–15 and Table 58)

4 Impact Assessment

4.1 Mainstream Hydropower Dams

This section estimates the impact of *avoiding* the construction and operation of 12 proposed lower Mekong mainstream hydropower dams. It is necessary to define the impact assessment question as a substitution of hydropower from the Reference

Table 4.4 Indicators, baseline, and reference scenario for the GMS energy system

Dimension/indicator	Metric/(unit)	Source/note	Baseline (2005 unless otherwise indicated)	Reference energy scenario (2025 unless otherwise indicated)
I Metabolism (Indicator 1) Primary energy demand	Change to primary energy demand (PJ)	IRM 2008	9913 PJ	26,665 PJ
II Governance (Indicator 2) Choice	Number of retail providers for grid electricity, per distribution network	Author; Nguyen (2012)	One	>1 (Vietnam) One (rest of Region)
(Indicator 3) Contention	Number of inter-annual disputes related to energy facilities[a]	Author	8	20
III Availability (Indicator 4) Hydropower supply	Share of hydroelectricity in regional electricity demand (%)	IRM (2008: Table 141)	20%	35%
(Indicator 5) Fossil energy resources and stockpiles	Years of remaining lowest-cost coal oil gas reserves divided by TPES	IRM (2008: 10–15)	176	50
IV Access (Indicator 6) Access to electricity in low access Mekong countries	Share of population in Cambodia and Myanmar electrified (%)	World Bank (2011), BAU scenario	26% Cambodia 13% Myanmar	40% (Cambodia) 20% (Myanmar)
(Indicator 7) Access to traditional biomass	Biomass share of regional primary energy supply (%)	IRM (2008: Table 142)	28%	15.6%
V Affordability (Indicator 8) Proportion of population in all Mekong countries unable to afford modern energy services	Maximum national poverty headcount at \$1.25 PPP (2005 USD) (%)[b]	World Bank (2012a)	22.8% (Cambodia 2008) 33.8% (Laos 2008)	0%
VI Technological development and long-Term sustainability (Indicator 9) Macro-economic energy efficiency	Energy consumption per dollar of GDP (MJ/year 2000 USD)	IRM 2008:250	31.4	24.9

(Indicator 10) Diversification to 'Nonconventional' renewable energy (NRE)	NRE in regional primary energy supply (%)[c]	IRM (2008: Table 142)	0.22%	3.5%
VII Resilience	Biomass power share of regional electricity supply (%)	IRM (2008: Table 141)	2.1%	8.4%
(Indicator 11) Decentralization				
VIII Environmental **Sustainability** (Indicator 12) Pollution	Thousand tons of SO2 (kt)	IRM (2008: Table 59)	1,922 kt	5,388 kt
(Indicator 13) Climate change	Million tons of CO_2 equivalent (mt)		523 mt	1,665 mt

Source: Author, based on above references

[a]Number of disputes related to bio-energy plantations, power plants, and associated infrastructure

[b]All countries excluding Myanmar

[c]Includes solar, wind, geothermal; does not include biomass

scenario, not an addition, because the Reference scenario already includes large increases in the region's hydroelectricity system relative to the volume of hydroelectricity from the 12 proposed dams.

In the Reference scenario, between 2005 and 2025, the share of hydroelectricity in the region's electricity supply increases from 19% to 35% (2–5.5% of regional primary energy supply), and the region requires a cumulative 5040 TWh of *hydroelectricity* (IRM 2008: Table 141).[4] By contrast, the 12 proposed dams have a combined capacity of 12,980 MW, and supply 64,229 GWh per year (not including projected flow increases from climate change). If we assume they become fully operational in 2016, the 12 dams would supply a total of 642.3 TWh during the 10 year period 2016–2025 (ICEM 2010). This equates to 12.7% of the additional hydroelectricity required during the 20 year period 2005–2025.

Assessment. Hydropower development allows conservation of other fuels, such as coal, natural gas, and/or deferral of investment in nuclear power. Nuclear power enters the Vietnamese and Thai power development plans in the year 2015 and 2020 respectively, however for Thailand at lower capacity and with greater uncertainty. Assuming that coal-fired power at 42% efficiency is a more feasible substitute for hydropower, 5505 PJ of coal is required to substitute for the 10 years of hydroelectricity from the 12 dams.

A more costly option would be to substitute combined cycle natural gas power plants for mainstream hydropower. Substituting mainstream hydro exclusively by gas-fired power at 50% efficiency would increase gas consumption by 4624 PJ natural gas between 2016 and 2025, approximately 50% of the region's lowest cost gas reserves. This could potentially raise natural gas prices, lowering the penetration rate of high efficiency combined heat and power systems for commercial buildings and combustion natural gas for transport, marginally increasing CO_2 emissions (*Indicator 13*) from both sectors.

Indirect energy requirements. Estimates of upstream and construction energy inputs per MW of capacity derived from life cycle analysis are shown in Table 4.5. For a 12,680 MW coal-fired power system, the total upstream energy requirement for construction, mineral extraction and transportation, and lifetime operation is approximately 1045 PJ (Table 4.5).

Assessment of life cycle energy requirements for hydropower construction and operation vary with scale of project and also system boundaries. Ribeiro and da Silva (2010) report a 15-fold lower ratio of energy input per MW capacity for Itaipu dam, compared to Spath et al.'s (1999) assessment of a low-emission coal plant. However the system boundaries of the latter study included energy requirements to construct mining machinery and transport vehicles such as coal barges.

Brazil's Itaipu dam at 14,000 MW exceeds the combined generating capacity of the 12 proposed Mekong dams, but as a single project it should have superior economies of scale. Assuming conservatively that building 12 Mekong dams incurs a factor of five higher lifecycle resource intensity (MJ energy/MW capacity) than

---

[4] ,and the vast majority of the power generated would be exported to Thailand and Vietnam.

Table 4.5 Indirect energy required per MW capacity, emissions, and energy return over energy invested

Technology	EROEI (MJ output/MJ input)	MJ energy input/MW capacity	PJ energy input for 12,680 MW	CO_2-equivalent emissions	SO_x emissions
Coal-fired power plant (42% efficiency, 60% capacity factor, 30 year lifetime)	6.7	8.24×10^7 (total non-coal energy)	1,044.8	20 g CO_2/ net kWh (mining and transport) 221 g CO_2/ net kWh (combustion of coal) 500 g CO_2/net kWh power (plant construction and operations)	0.46 mt total 0.715 g/kWh (lifecycle)
Hydropower (Dhukwan 30 MW)	25	3.04×10^7	385.5		
Hydropower (Itaipu 14,000 MW)	403.6	5.31×10^6	67.3	0.93 mt (construction) 1.08×10^{-3} t/ MWh (operation)	0.0013 mt (construction) 0 mt (operation)

Source: Author, based on life cycle analyses for coal (Spath et al. 1999; Harvey 2010b); Itaipu hydropower (Ribeiro and da Silva 2010); Dhukwan hydropower (Varun and Bhat 2010)

reported for Itaipu, the construction and operation energy requirement is $(5 \times 67.3) = 337$ PJ for the Mekong dams. The *net* incremental non-coal energy requirement of substituting coal plant for these dams is approximately 708 PJ.[5]

Adding 5505 PJ of coal energy results in a total net energy requirement of +6213 PJ, increasing regional energy demand by 2.9% between 2016 and 2025. Higher indirect energy requirements marginally lowers remaining fossil energy reserves (*Indicator 5*) and raises energy intensity (*Indicator 9*). In terms of carbon emissions (*Indicator 13*), the coal-based system emits 741 g CO_2/net kWh electricity produced, with the following attribution:

20 g CO_2/net kWh emitted during mining and transport;
221 g CO_2/net kWh emitted from combustion of coal;
500 g CO_2/net kWh emitted by power plant construction and operations (Spath et al. 1998: Table 29; IRM 2008: Table 27)

[5] These estimates overstate actual impacts because they include (for consistency) maintenance energy requirements for the entire lifecycle of each technology. However in the case of Itaipu dam including 10 vs. 100 years of maintenance requirements lowers PJ input/MW by only 15%.

Using these parameters, the total emissions from a coal-based system producing 642.3 TWh is approximately 188 mt CO_2 equivalent (of which 76% is from combustion of coal, 17% from power plant construction and operations, and 7% from mining and transport).

By contrast, the CO_2 emissions required to generate 642.3 TWh from the Itaipu dam are only 1.62 mt (0.93 mt for construction, 0.69 t for operations; Table 4.5). Assuming construction and operation of the Mekong dams produces five times the carbon emissions per kWh than does Itaipu, the net incremental emissions from avoiding hydropower are approximately 188–8 = 180 mt. Similar calculations for sulphur emissions result in net additional emissions of approximately 0.46 mt. Hydropower because of its faster ramp rate allows better integration of intermittent renewable energy (e.g. wind and solar) than does higher cost gas-fired power (Harvey 2010a). Avoiding hydropower could marginally lower the pace of diversification to non-conventional RE *(Indicator 10)* compared to the Reference scenario.[6]

Levels of contention *(Indicator 3)* are unchanged from the Reference scenario, where they rise as a result of the increasing number of power plants required by a growing system. The 12 mainstream dams are located far from most population centres. Coal or natural gas plants require sources of cooling water and access to resources via port or pipeline, and thus areas near infrastructure are preferred siting locations. In Thailand, all large power plants are to some degree controversial, with coal arguably more than natural gas, and both more contentious than border hydropower. Contention will rise if coal plants shift into Thailand closer to population centres. Investors know this and may be able to avoid siting disputes by situating fossil power plants outside Thailand but adjacent to the border (e.g., Koh Kong in Cambodia), thus this indicator is assessed as unchanged.

4.2 Water Diversions from Laos into Northeast Thailand

Proposals have been made since at least the 1990s to divert water either from the Mekong or from Lao tributary rivers into *Isan*, the large Northeast region of Thailand which has long been framed poor, dry, and politically sensitive. They are consistent with grand and longstanding hydraulic engineering approaches to developing the Mekong region itself promoted by the United States during the Cold War (Molle et al. 2009b).

At least two Lao-Thai diversion proposals exist: from the Nam Ngum Basin near Vientiane, and from the Xe Bang Fai Basin in central Laos. According to Pech (see Chap. 2 in this volume) the Nam Ngum scheme would involve an annual 300 m^3/s (approx. 6,880 mcm/year) water transfer piped under the Mekong to feed into

[6] This assumes that the IRM model has not accounted for such synergies.

existing water storages in the Chi-Mun basin in Thailand (Floch and Blake 2010). Alternatively, with agreement from the three other 1995 Mekong Agreement signatories (Laos, Cambodia, and Vietnam) Thailand could divert water directly from the Mekong into Thailand's Loei province.

There is a lack of publicly available information, and no prior international water transfer scheme exists in the region. The current proposal however is in a series of developmental proposals for Lao-Thai water resource sharing, driven by visions of broad-based rural development enabled by water resources infrastructure (Molle et al. 2009a; Floch and Blake 2010).

Here we assess the potential impacts of diverting up to 6,880 mcm of Mekong basin water to irrigate dry-season crops in Thailand. Specifically, we estimate the impact of *avoiding* the production of ethanol from irrigated sugar cane that could be grown in Northeast Thailand using varying amounts of up to 6,880 million cubic meter/year of water imported from the Mekong or from Laos tributaries. It is necessary to define the impact assessment question as a substitution of biofuel from the Reference scenario, not an addition, because the Reference scenario already includes significant increases in biofuel production relative to the volume of ethanol that could be produced with this water allowance.

Total ethanol production would increase if the water transfer resulted in increased usage of irrigation water to expand sugar cane cultivation from its rainfed base.[7] Based on Thai irrigated land expansion scenarios, land availability is not limiting.

Assessment: If ethanol were *not* produced from expanded irrigated sugar cane in NE Thailand, then 519 million L/year of combustion gasoline would need to be substitute (18.4 PJ crude oil/year), leading to marginal declines in diversification to renewables (*Indicator 10*). Economic energy intensity *(Indicator 9)* would also increase marginally. The economic energy intensity of ethanol production under alternative practice conditions is 9.7 MJ/2,000 USD. However, the additional 184 PJ of crude oil substituted for ethanol over the period 2016–2025 increases GHG emissions and lowers the supply of lowest-cost fossil energy reserves only marginally (by +3.7 mt and −0.07 years respectively).

Ethanol yields for three cases based on alternate assumptions about the fraction of imported water that could be applied to an irrigated sugar cane crop are shown in Table 4.6. Conventional production of ethanol requires a net energy input of 27 MJ/L (of which 15.2 MJ/L are fossil fuel inputs). The alternative practice case is to use 50% of bagasse produced (cane trash) and 100% of the stillage produced (a by-product of distillation) to supply part of the refinery's energy requirement, lowering net energy input to 7.3 MJ/L of which only 2.93 MJ/L are fossil inputs (Nguyen et al. 2008: 732).

[7] Yields per ha could also increase. While Thailand's cassava per ha yields are among the highest in the region, and better than the rest of Asia, its sugar cane yields are less than other producing regions (Malik 2009; Takeshita 2009).

Table 4.6 Potential ethanol yields as a function of expanded irrigated sugar cane production, alternative practice case

Case	1	2	3	Source
Assumed volume diverted	6,880 mcm per year			
Fraction assumed used for sugar cane	90%	50%	10%	
Blue and grey water requirement (m^3 per ton sugarcane)	60	60	60	Estimated from Gerbens-Leenes and Hoekstra (2012)
Sugarcane yield t	1.03E+08	5.73E+07	1.15E+07	
L ethanol per t cane (molasses path)	10.18	10.18	10.18	Nguyen et al. (2008)
L ethanol	1.05E+09	5.84E+08	1.17E+08	
Million L	**1,051**	**584**	**117**	
Million L per day	**2.88**	**1.60**	**0.32**	
As % of 2,021 Thai government target (9 million L/d)	**32%**	**18%**	**4%**	
Annual energy yield (PJ) (energy balance)	**33.1**	**18.4**	**3.7**	
As % of GMS biofuel supply, reference scenario	**158%**	**88%**	**18%**	IRM (2008: Table 140)
Land requirement (ha)	9.05E+06	5.03E+06	1.01E+06	Nguyen et al. 2008
As % of Thai irrigated area expansion scenarios	156%	87%	17%	Chayanis K. pers. Commun.

Source: author, based on above references

Assuming 3,440 mcm/year water (Case 2), and assuming the alternative practice case, net renewable energy outputs are 584 million L/year of ethanol, or 18.4 PJ.[8] This is 88% of the Reference scenario and 18% of the Thai government's more ambitious ethanol production target for 2021.

By contrast, a scenario where production of ethanol from irrigated sugar cane is *avoided* may marginally *improve* affordability (*Indicator 8*) based on analysis conducted by Salvatore and Damen (2010). First, Thailand is assumed to be a price taker, and increases in world ethanol prices are assumed to be eventually transmitted to domestic consumers in the form of higher food prices. This increases farming incomes for all but the poorest farming households. Simulated through a price shock

[8] Calculated using the energy balance (EnB) approach: based on comparative fuel economy values, 1 L of ethanol substitutes for 0.89 L of combustion gasoline (CG). One L of CG has 38.7 MJ of primary energy. The result of 31.5 MJ/L net avoided fossil energy is equal to (0.89×38.7), less 2.93 MJ/L net fossil energy inputs to ethanol production (Nguyen et al. 2007: Table 6, Scenario 3).

effect in a CGE model, a 1% increase in food prices decreases household incomes for all non-farm households, with the poorest category of non-farm households incurring a −1.6% decline in real income (Salvatore and Damen 2010: Table 8.9). We would expect immigrant households from Myanmar and Cambodia to be in the lowest income categories of non-farm households in Thailand. A separate study by ADB (2011) based on similar methods estimated that a 10% increase in food prices would increase the $1.25 PPP poverty headcount by 1.9% in developing Asia with the headcount rising from 27.1% to 29% (ADB 2011: 27).

In the long term, if Thai farmers could increase sugar cane yields, or if ethanol production efficiencies improve, then higher output per unit input could lower prices (Salvatore and Damen 2010). If Thailand remains a price taker it is lower feedstock or production prices might not be passed on to consumers. If not, then affordability would not improve. Market prices of Thai sugar cane ethanol-per gasoline equivalent litre have typically been significantly higher than gasoline, requiring ongoing subsidies (Nguyen et al. 2007).

Plausiblity. The importance of irrigation water as the most important factor limiting crop production in NE Thailand however needs to be questioned (Molle and Floch 2008). Returns to farming are influenced not just by water availability, but the profitability of smallholder farming, taking into account a range of output *and* input prices. High crop prices might increase returns to farming and justify investment in more intensive farming systems. But an increase in crop prices may also be accompanied by an offsetting increase in fertilizer and other input prices. Those chemical inputs may sensitive to possible increases in fossil fuel prices. Molle and Floch (2008) note that agricultural expansion in NE Thailand must contend with the problem of saline soils and long-term declining rice prices. They argue that dry-season agriculture in Thailand is limited by shortage of on-farm *labor*, not water. A study on farming practices in four districts in the lower Mun basin found that pumped water was used primarily to raise rice seedlings for the main rain-fed rice crop, and secondly to grow higher-value crops such as watermelon and chilli, as well as stock fish ponds in the dry season. In 2000–02, the average actual usage rate was only 14% of the total estimated irrigable area (UBU 2002: *Khor*-6). The main obstacle to expanded use of pumped water was the high user charge relative to the value of rice (Foran 2006: ch. 8). Other authors cite a similar irrigated area utilization rate of 14% (Kamkongsak and Law 2001).

4.3 Sea-Level Rise and Adaptation Strategies in Vietnam's Mekong Delta

Description: Warmer global temperatures lead to a 20 cm sea level rise by 2030, rising to 100 by 2,100, exacerbating disruptive high tide events, accompanied by possible extreme weather events, threatening viability of human settlement in portions of the Mekong Delta and beyond. Under this scenario decision makers decide to invest in structural measures to defend 85% of the Mekong Delta from inundation, and 15% of the present area is lost by 2100 (see Chap. 1 this volume),

Assessment: A direct effect on the energy system is to increase the vulnerability of energy infrastructure (ports, refineries, power stations) to inundation, erosion, and extreme weather events. Adapting to a 20 cm sea level rise and associated climate change by 2030[9] would impact on the energy system in several ways (Agrawala and Fankhauser 2008), notably:

1. Demand for materials and energy to fortify, maintain, or relocate refineries, power stations and other energy infrastructure;
2. Similarly, demand for materials and energy to fortify, maintain or relocate non-energy infrastructure;
3. Increased demand for space conditioning resulting from warmer temperatures;
4. Increased demand for energy required as an input to water and sanitation systems (e.g. desalination plants);
5. Impacts on energy services and prices resulting from behavioral, social, planning, and other institutional responses (e.g. migration, regulatory changes)

This section focuses on the first two types of impact. We assume engineering adaptation responses take place over 10 years, from 2016 to 2025. Considerable uncertainty exists with respect to adaptation costs and the study of adaptation options and costs is still in its infancy (Fankhauser 2010). Empirical estimates of the cost of coastal protection vary by region but are less than 0.5% of GNP for all but the world's smallest states or regions (Agrawala and Fankhauser 2008: Table 2.3).

Assuming the incremental cost of adaptive responses to sea level rise ranges from 0.1% to 0.5% of regional GDP (undiscounted), primary energy demand 2016–2025 increases linearly, by 218 PJ to 1088 PJ. This simple case assumes no change to Reference scenario trajectories of energy intensity (MJ/$ GDP) which declines by 1.4% during this period. A more precise assessment would use specific analysis of planned adaptations to infrastructure, to derive changes in demand for cement, iron and steel, construction equipment, fuels, and energy-intensive inputs.

Economic energy intensity (*Indicator 9*) is expected to decline. It could worsen if adaptation requirements compete for investment capital, leading to lower economic activity than foreseen in the Reference scenario. Maintaining an adequate flow of investment in countries like Vietnam would require tighter fiscal policies and corporate governance (ADB 2011: 217).

Affordability (*Indicator 8*) declines as the following example illustrates. Assume that engineering adaptations to sea level rise result in a 10% increase in the investment cost (USD/kW) for a coal or natural gas fired power plant located in a vulnerable area. Assume that coal and natural gas fuel costs also increase by 10% to cover the cost of fortifying ports, pipelines, and associated infrastructure. The combined impact of a 15% increase in investment cost and a 10% increase in fuel cost is to increase the levelized cost of electricity (USD/kWh) by 7.6% for coal-fired power

[9] The effects of 100 cm rise would be severe for the Mekong and Chao Phraya Delta, and also extend to energy plant concentrated on Thailand's Eastern Seaboard (Gulf of Thailand). Strategic energy hubs beyond the region (Singapore) and other coastal cities would also be vulnerable.

and 8.5% for gas-fired power. Similarly, the price of petroleum products from oil refineries incurring similar capital and fuel cost increases could rise by a comparable amount. Higher energy prices lower economic growth and increase general inflation, both of which reduce the incomes of poor households (Asian Development Bank 2011). Contention (*Indicator 3*) increases as communities in a context of volatile and possibly lower average economic growth dispute distributional impacts of the infrastructural investments (cf. contention observed north of Bangkok over plans to fortify industrial estates after the 2011 floods).

4.4 Rubber Plantations

Description: More than 5,000 km^2 of upland forest in montane mainland SE Asia may have been converted to rubber, and the area dedicated to 'rubber and other diversified farming systems' – that is, to permanent agriculture – may double or triple by 2050 (Ziegler et al. 2009b). Upland areas of the GMS (elevation 250–3,000 m) cover 55% of the region's area, with 31% of its population, mostly living in Yunnan (Johnston et al. 2010). Johnston et al. divide the uplands into 'intensively used' (100–250 persons/km^2, 10% of GMS) and 'forested' categories (<50 persons/km^2, 45% of GMS area). Since the total upland area in the GMS is 1.6 million km^2 (Johnston et al. 2010) an increase to 15,000 km^2 would represent less than a 1% conversion of the uplands. Nothwithstanding the small fraction of upland area implied, several analysts draw attention to immediate negative impacts on biodiversity, with loss of primary and secondary forest. In the context of montane or upland regions, they also point to negative impacts such as soil erosion, stream dessication in the dry season, and water pollution from the use of pesticides and fertilizers (Ziegler et al. 2009a, b).

Potential impacts on energy system: We expect marginal increases in energy demand (*Indicator 1*) for agricultural inputs associated with upland farming expansion. Access to traditional biomass (*Indicator 6B*) may well decline if concessions lead to dispossession. Impacts such as affordability (*Indicator 7*) are difficult to quantify in the absence of more detailed studies. Decentralization (*Indicator 10*) may increase, since some portion of the permanent agriculture that is expanding in the region will include crops such as maize, cassava, timber, and jatropha. Residues production could contribute to community-scale energy systems. Old rubber trees are also source of biomass energy, albeit with a rotation period of 25–30 years (Abe 2007).

Ziegler et al. (2009a) suggest that intensified agriculture can reduce streamflow in the dry season, and increase surface erosion leading to increased stream sediment loads. If catchments were converted to rubber plantations at significant scale, in the absence of effective upland management practice, a potential exists for losses to hydropower production. Such impacts have not yet been reported but might emerge first for hydropower projects in small catchments. But assuming regional climate models are accurate, rubber impacts on regional hydropower production should be more than offset by higher precipitation (Table 4.3).[10]

4.5 Developing Rail Infrastructure

Background: The GMS countries and territories could be linked by a number of alternate railway routes. So-called Route 1 would link Bangkok, Phnom Penh, and Ho Chi Minh City, connecting to Hanoi via Vietnam's north–south rail system, then to Kunming and Nanning. Including the Vientiane to Bangkok connection, the total distance along this route is 4,537 km (Canarail Consultants 2011). Route 1 would pass through all countries except Myanmar, and had the highest economic internal rate of return (75% EIRR) and 'marketability' according to Canarail (2011). Alternatively, Route 2 involves linking Thailand and Laos to Yunnan in the form of a rail project from Vientiane northwards to Xishuangbanna in southern Yunnan, then to Kunming. Route 2 would require a minimum of 1,092 km of new railway at a cost of $5.32 billion, with a 24% EIRR (see Fig. 4.6).

Route 1 involves significantly less new railway construction (397 km at 1.1 billion USD), however countries such as Vietnam are planning to expand capacity of existing lines, e.g. planning a high-speed electric passenger connection between Hanoi and Ho Chi Minh city, which would require additional trackway (Canarail 2011: 12). For the purpose of this assessment we focus on Vientiane-Bangkok-Kunming via Route 1 and assume 4,537 km of new rail system is constructed.

Assumptions for impact assessment: We focus on a first-order conservative estimate of the impacts of diverting passenger travel (by bus and air) and freight travel (by truck) to electric rail, Route 1. Construction and upgrading of rail networks between the GMS members allows the mode share of electric rail to increase against freight and passenger transport by alternative modes, such as bus, truck, and air. Figure 4.7 shows projected rail traffic on Route 1 based on estimates by Canarail (2011) of new travel and mode shift to rail. Figure 4.8 shows estimates of annual primary energy requirements for those volumes of traffic. These estimates include energy required for vehicle operation, as well as system maintenance and infrastructure operation, including energy to administer human resources and liability services (Chester and Horvath 2010).

Indirect energy requirements. Energy system impacts extend upstream beyond direct energy use per passenger-km (p-km) or ton-km (t-km). They include energy required for constructing trackway, vehicles, stations, and high embodied-energy tunnels and bridges. Additional energy is required for system maintenance. Energy requirements depend on a number of system-specific logistical, technical, and operational parameters (Van Wee et al. 2005). Thus in addition to operational energy requirements, it is important to account for indirect (embodied) energy requirements (Chester and Horvath 2009).

Route 1 requires a minimum of 397 km of railway. In addition, we assume conservatively that a new electric passenger and freight service will require 4,140 km of new trackway (equal to the existing 4,140 km diesel network). To estimate indirect

[10] Beyond the time scale of this assessment land use change to monocultural rubber could contribute to increased sediment input to reservoirs (Johnston et al. 2010), lowering energy production and raising costs marginally.

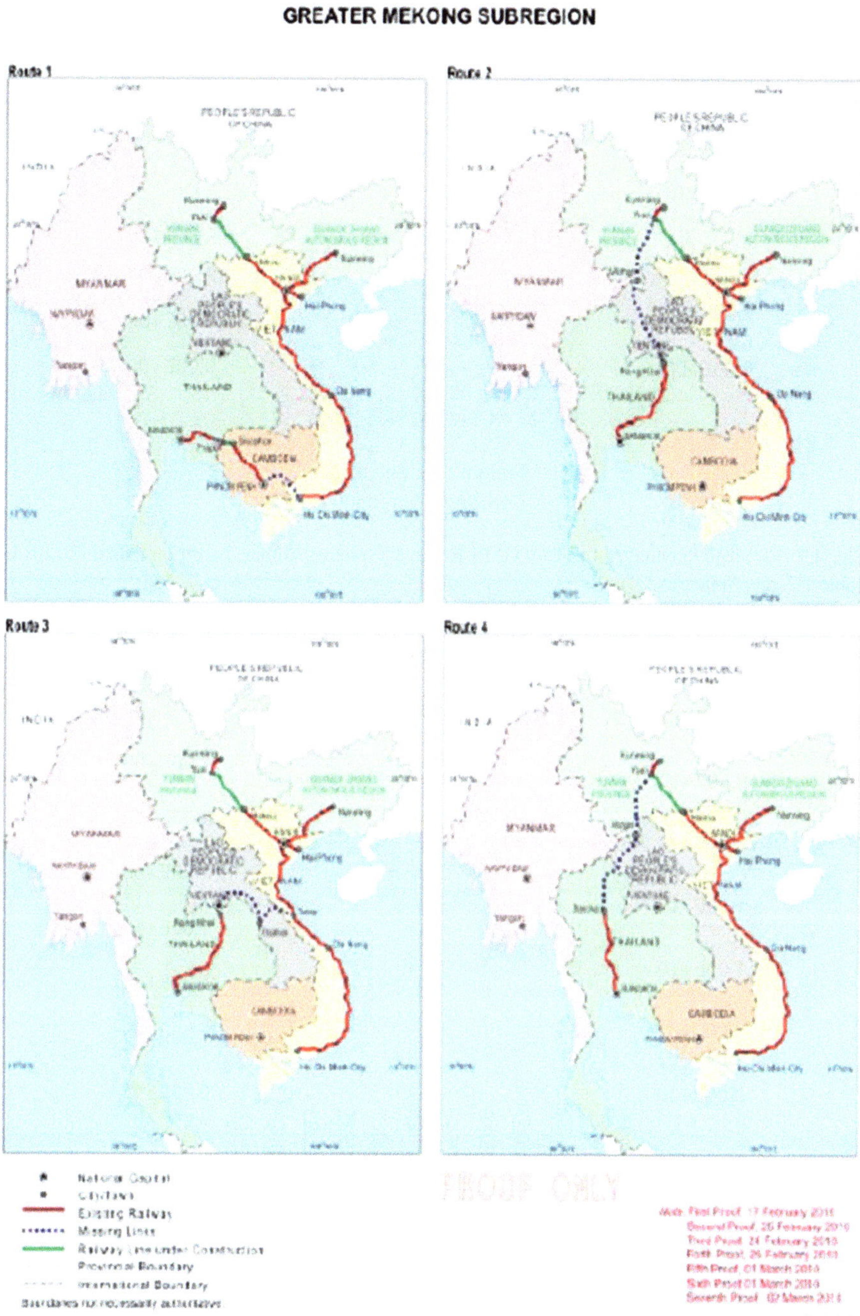

Fig. 4.6 Four primary GMS rail routes (*Source*: Canarail 2011)

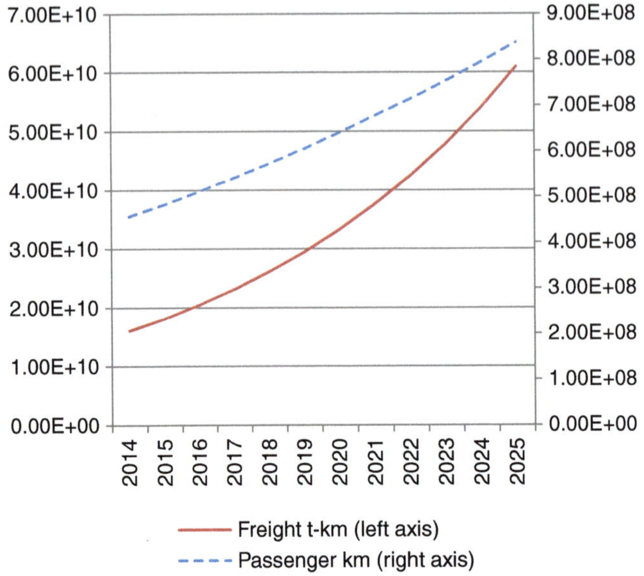

Fig. 4.7 Estimated annual rail traffic, GMS Route 1 (*Source*: Author, based on Canarail (2011: Table 7.7 and Appendix C))

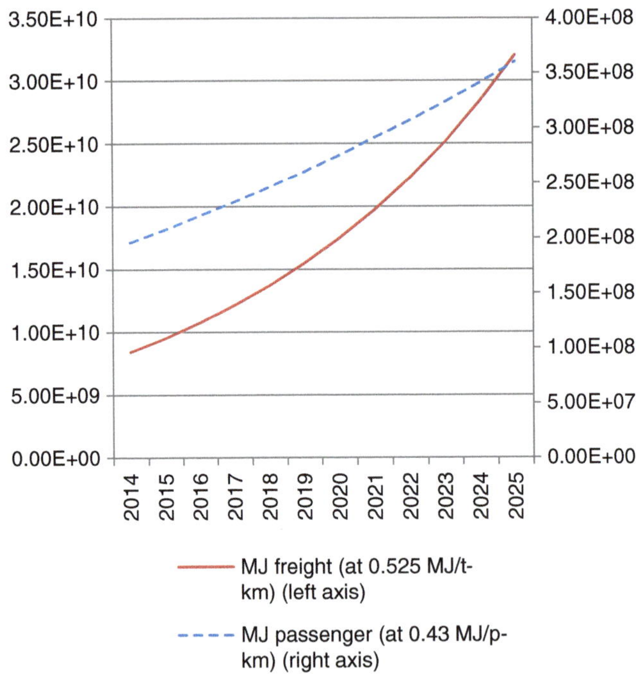

Fig. 4.8 Annual requirements for primary energy, GMS Route 1 (MJ) (*Source*: Author, based on vehicle operation requirements in Harvey (2010a: Table 5.3), assuming primary energy markup factor of 2.5 for electricity, plus 5% additional system operation and maintenance energy requirements (Chester and Horvath 2010))

energy requirements in the absence of GMS-specific rail network parameter estimates, we drew on material (kg/km-year) and energy intensities (MJ/kg) provided by Federici et al. (2008), who conducted an embodied energy analysis of the Inter-City Milano-Napoli electric rail line.[11] Results for GMS Route 1 are shown in Table 4.7.

To the extent that rail vehicles and other system components are *not* produced within the region, and to the extent that existing civil works and system components can be used, we have overstated the indirect energy requirements. A second conservative assumption is that we have not subtracted the avoided energy requirements for road and air travel and infrastructure.[12]

Potential effects on energy system: For the period 2014–2025, annual energy requirements add up to 218 PJ, or 2.1% of the region's final energy demand for freight transport final in the IRM Base scenario (2008: Table 56). Indirect requirements amount to 1298 PJ.

Assuming the IRM (2008) simulations have already subsumed the annual requirements, but not accounted for the indirect requirements, the construction and renovation of Route 1 as an electric rail network leads to more notable increases in primary energy demand. The magnitude of the increase depends on the period of time construction is assumed to take place, e.g. regional primary energy demand increases by 3% if Route 1 is completed 2011–13, or 0.8% if assumed to require 10 years 2011–2020 (*Indicator 1*).

For the general rural population, affordability (Dimension V) of fuels such as diesel and LPG for cooking may increase slightly in rural areas with new or enhanced access to rail service. The average cost per t-km of general freight transported by road is as high as 12 times the cost of rail transport (Harvey 2010a, Table 5.30). The cost of transporting oil is higher than general freight, approximately 2.4 times for Thailand-Laos (Canarail 2011). Table 4.8 estimates potential cost savings on remote areas in the GMS. Estimates are based on road freight charges estimated for southern Laos for general goods (Asian Development Bank 2005), and rail charges for Nong Khai-Vientiane (general goods and oil; Canarail 2011: 58).

Notwithstanding these potential impacts on petroleum product prices in remote areas, we expect that any change in affordability for the very poor (*Indicator 8*) would require rail development to raise incomes, rather than merely lowering energy expenditures. For example, construction of the railway could induce the growth of existing and new settlements, leading to increased informal employment opportunities. Analysts have concluded however that providing initial road access to isolated households in Laos contributes more to poverty reduction than upgrading road connections to households already serviced (Brooks 2008). Thus we assess the impact on affordability as only a marginal increase.

---

[11] In 2001, this 778 line carried approximately 4.55×10^9 t-km of traffic, slightly less than one third of the traffic estimated for GMS Route 1 in 2014, its first feasible year of operation.

[12] Doing so would yield a more precise answer but would also require more detailed information on traffic shifted to rail from each mode.

Table 4.7 Estimated indirect energy requirement for construction of GMS Route 1

			GMS Route 1 (Vientiane-Bangkok-Phnom Penh-Ho Chi Minh City-Hanoi; Hanoi-Kunming; Hanoi-Nanning)		
Items	Lifetime (year)	Material intensity (kg/km-yr)	Length (km)	Materials (kg)	Energy (MJ)
Sand and gravel	50	6.96E+06	4,537	1.58E+12	1.58E+10
Moved soil	50	1.94E+06	4,537	4.40E+11	(covered by diesel)
Concrete	50	7.20E+05	4,537	1.63E+11	7.52E+11
Reinforced concrete	50	6.91E+04	4,537	1.57E+10	4.25E+11
Steel in tunnel reinforcement	50	4.54E+04	60	1.36E+08	1.09E+10
Diesel for civil works	1	2.20E+03	4,537	9.98E+06	5.33E+08
Steel in construction machineries	30	3.59E+00	4,537	4.89E+05	3.91E+07
Steel in track	10	2.34E+04	4,537	1.06E+09	8.49E+10
Steel in electric poles	10	1.36E+02	4,537	6.17E+06	4.94E+08
Copper in electric cables	10	4.86E+02	4,537	2.20E+07	2.92E+09
Steel in passenger trains (n=96)	30			5.20E+07	4.15E+09
Steel in freight trains (n=30)	30			1.30E+07	1.04E+09
Steel in maintenance vehicles	30			1.10E+06	8.79E+07
				Total	1,298 PJ

Source: Author, based on Federici et al. (2008: Tables 2, 5)

Table 4.8 Estimated potential cost savings of transporting petroleum products by rail

Distance (km)	Oil by rail (USD/L)	Oil by road (USD/L)	Potential savings (USD/L)
100	0.009	0.022	0.013
200	0.019	0.045	0.026
300	0.028	0.067	0.039
400	0.037	0.090	0.052
500	0.047	0.112	0.065
600	0.056	0.135	0.079

Source: Author

Notes: Oil by rail $0.11/t-km; oil by road assumed $0.264/t-km based on Canarail (2011) and ADB (2005: Table A6.3)

4.6 Bauxite Mining and Aluminium Production

Description: We explore impacts on the energy system assuming the region produces 1 million tons per year of alumina beginning in 2016 from potential new bauxite mines in the region, notably located in southern Lao PDR. We also explore the energy impacts of further value-adding by producing primary aluminium ingot.

Approximately 2.15 ton of dry bauxite are required per ton alumina, and 1.95 ton alumina is required per ton primary aluminium (i.e., 4.2 tons of bauxite per ton aluminium) (Harvey 2010a: 352). Table 4.9 shows electricity and fuel energy requirements for mining bauxite, refining to alumina, and smelting of primary aluminium (i.e. aluminium that is not mixed with post-consumer scrap). Fuels required are for heating processes and usually sourced from natural gas or coal. Smelting of primary aluminium from alumina is energy intensive: approximately 1,190 MW electricity is required to smelt one million tons of alumina per annum. Competitive smelting requires reliable electricity supplies costing US 2.5 cents per kWh, compared to estimated generating costs in Lao PDR of 5–7 cents (Larsen 2010).

Potential impacts on energy system: Direct energy requirements to produce 1×10^6 tons of alumina are approximately 17 PJ p.a., or 85.4 PJ per annum to produce 513,000 t aluminium. Annual *indirect* energy requirements include energy to transport bauxite to a refining/smelting complex, energy to extract and transport coal (the default fuel) to the complex. Table 4.10 below provides an estimate of all annual energy requirements, based on input–output analysis of the U.S. economy by Carnegie Mellon University (2012). Total annual energy requirements for all sectors of the US economy that supplied inputs to alumina refining and primary aluminium production in 2002 is shown, along with energy requirements of the top 10 sectors. Assuming the requirements of sectors 1–3, 7 and 10 (petroleum refining) have already been accounted for in Table 4.9, the energy requirement of all remaining sectors is 8.4% of the total.

Assuming that the annual indirect requirements of Mekong aluminium production do not exceed 108.4% of the direct energy requirements, the total annual energy requirement to smelt 1 million tons of alumina (into 513 kt aluminum) is 92.6 PJ/annum.

Energy to build new refining and smelting plant and infrastructure complexes cannot be estimated from input–output data. Assuming that those initial requirements do not exceed 300% of the total annual energy requirements, the incremental impact on regional energy primary energy demand of 10 years' start-up production of *aluminium* is 278+ 926 = 1204 PJ. If only *alumina* is refined the corresponding impact of 10 years' production is 222 PJ.

Trade-offs with biomass consumption are expected for areas impacted by mining and development. Losses to hydropower production may occur in some areas if streamflow is abstracted for alumina refining (Lazarus 2009:23). Alternatively, producers may invest in hydropower plant construction, allocating water resources in a river basin so as to minimize negative trade-offs between alumina and hydropower production.

Table 4.9 Onsite energy intensities and energy requirement to produce bauxite, alumina, and aluminium

	Final energy (GJ/t)		Primary energy (GJ/t)			PJ (1 million t/a alumina or 513,000 t/a Al)		
	Electricity (kWh/t)	Fuel (GJ/t)	40% efficiency	Reference scenario (53% efficiency)	90% efficiency	40% efficiency	Reference scenario (53% efficiency)	90% efficiency
Mining (product = bauxite)	15	0.05	0.20	0.16	0.12	0.42	0.35	0.26
Refining (product = alumina)	250	12.50	17.3	16.7	16.0	17.25	16.7	16.00
Smelting (product = Al)	16,000	18.00	166	133.2	86	84.92	67.3	43.90
Total (Alumina)	265	12.55	**17**	**17**	**16**	**17.7**	**17.0**	**16.26**
Total (primary aluminium)	16,265	30.55	**183**	**150**	**102**	**102.6**	**85.4**	**60.16**

Source: Author, based on Harvey (2010a: Table 6.7)
Notes: 'Fuel' assumed to be coal. Primary energy of fuel assumes markup factor of 1.2. 'Efficiency' refers to efficiency of electricity generation and transmission; Reference scenario (IRM 2008: Table 120); 40% approximates coal-dominated generation; 90% hydro-dominated generation

Table 4.10 Primary aluminium, annual indirect energy requirements (GJ/t)

		GJ/t Aluminium	%
	Total for all sectors	**70.15**	100.0
1	*Alumina refining and primary aluminium production*	37.37	**53.3**
2	*Power generation and supply*	21.48	**30.6**
3	*Secondary smelting and alloying of aluminium*	4.61	**6.6**
4	Truck transportation	0.75	1.1
5	Aluminium product manufacturing from purchased aluminium	0.50	0.7
6	Iron and steel mills	0.49	0.7
7	*Oil and gas extraction*	0.44	**0.6**
8	Rail transportation	0.40	0.6
9	Pipeline transportation	0.38	0.5
10	*Petroleum refineries*	0.36	**0.5**
	All sectors excluding sectors 1–3, 7, 10	**6.35**	**8.4**

Source: Author, based on Carnegie Mellon University, EIO-LCA database (2012)

5 Cumulative Impacts

What higher-order impacts might occur to the Mekong region's energy system if all six development impacts assessed above occurred in the near future? Such a scenario is admittedly implausible. It would be more realistic to think in terms of approvals taking place in the near future, followed by a necessary time lag as plans are revised, as markets and regulatory systems respond, as capacity and capital is harnessed. Only through such a phased series of actions and reactions is it possible to imagine all of the development occurring.

The final column in Table 4.11 summarizes cumulative impacts on the region's energy system to 2025 based on the impacts defined in Sect. 5. Table 4.11 shows notable increases in total primary energy demand (+5%), declines in Grade 1 fossil reserves (−6%), and declines in hydroelectricity share of total electricity (−7%). Emissions are expected to increase by at least 3%. These changes appear minor, but are notable in the sense that they reinforce, as opposed to countervail, prevailing system trajectories.

In this assessment, the poverty headcount at $1.25 per person per day (Sect. 1) is our metric for the fraction of the population unable to afford modern energy services. If regional GDP grows at 6.3% as assumed in the Reference scenario, the headcount falls to zero in all countries except Myanmar (for which data not available). However, rising energy prices impact on food and other prices and thus strongly on poverty as discussed in Sect. 3 of Chap. 5 (ADB 2011). In this assessment, the most notable impact on energy prices occurs from adaptation responses to sea-level rise and associated climate impacts on ports and coastal cities. We showed how infrastructural investments could plausibly trigger additional regional energy prices increases of 8–9% above business-as-usual dynamics.

The irrigation expansion assessment (Sect. 3, Chap. 5) focussed on the marginal impact of ethanol production using 3,440 mcm of water (50% of a potential 6,880 mcm diversion to Thailand). If irrigated agriculture becomes more viable in Northeast Thailand (cf. Molle and Floch 2008) a potential exists for expanded renewable power production fuelled by rice husks and other agricultural residues. More biomass power would allow Thailand to avoid some portion of either its hydropower or its bituminous coal imports, 'freeing' up hydropower for industrial development in Laos. A potential exists for long-term linkages between hydropower development and aluminium smelting, particularly as the cost of hydropower declines (after capital intensive hydro projects pay off their construction costs). However, in a context of higher energy and input prices, additional use of irrigation water from diversions to Thailand would require well-resourced farmers growing high-value added crops.

In the absence of climate change and other infrastructure developments, a 300 m^3/s diversion upstream of any proposed mainstream dam could lower power production linearly (by a factor of 300/plant design discharge). For example a diversion of 300 m^3/s to Thailand's Loei Province, in the upstream vicinity of the proposed Pak Chom hydropower project (designed to discharge up to 5,720 m^3/s), represents a potential loss of 5.2% of power production in the wet season, and more

Table 4.11 Summary of impacts of six regional developments on Mekong energy system

Dimension: # *Indicator*: metric (unit)	Baseline (year 2005 unless otherwise indicated)	Reference energy scenario (year 2025 unless otherwise indicated)	Avoiding mainstream hydro	Avoiding ethanol produced by water diversion	Sea level adaptations	Rubber	Railways	Alumina/ aluminium	Cumulative (2016–2025)
I Energy metabolism: *(1) Primary energy demand*: additional primary energy demand (PJ)	9913 PJ	26,665 PJ 293,446 PJ (2011–25) 217,600 PJ (2016–25)	+6213 PJ (2016–25)	+184 PJ (2016–25)	+218 to +1088 PJ (2016–25)	Marg inc	+1298 PJ (2011–2013)	+222 PJ (Alumina) +1204 PJ (Aluminium) (2016–25)	+8043 to +9895 PJ +3% (2011–25)
II Governance: *(2) Choice*: Number of retail providers for grid electricity, per distribution network	1	>1 (Vietnam) 1 (rest of Region)	NC	NC	NC	NC	NC	NC	NC
(3) Contention: number of disputes related to siting of energy and related facilities	8	20	NC	NC	Increase	NC	NC	NC	>20
III Availability: *(4) Years remaining Grade 1 coal oil gas reserves*	176 y	50 y	£−2.7 y 5.4%	Marginal decrease	−0.37y −0.7%	Marginal decrease	−0.07y 0.14%	−0.3y 0.7%	−3.2 y −6.4%

(5) Share of hydroelectricity (all GMS sources) in regional electricity demand (%)	19.5%	35%	−7% (2016–2025)	NC	NC	Marg decrease	NC	Marg increase	−7%
IV Access: (6) *Access to electricity in low access Mekong countries*: share of population in Cambodia and Myanmar electrified (%)	26% Cambodia 13% Myanmar	40% (Cambodia) 20% (Myanmar)	NC	NC	NC	NC	NC	NC	NC
(7) Access to traditional biomass: biomass share of regional primary energy supply (%)	28%	15.6%	NC	Marg decrease	NC	Marg decrease	NC	Marginal decrease	Marginal decrease

(continued)

Table 4.11 (continued)

Dimension: # *Indicator*: metric (unit)	Baseline (year 2005 unless otherwise indicated)	Reference energy scenario (year 2025 unless otherwise indicated)	Avoiding mainstream hydro	Avoiding ethanol produced by water diversion	Sea level adaptations	Rubber	Railways	Alumina/ aluminium	Cumulative (2016–2025)
V Affordability: *(8) Proportion of population unable to afford modern energy services*: poverty headcount at $1.25 PPP (2005 USD) (%)	22.8%	0%	NC	Marg increase	Increase	NC	NC	NC	Increase
VI Long-term sustainability of supply: (9) *Energy intensity*: energy consumption per dollar of GDP (MJ/year 2000 USD)	31	25	Marg increase	Marg decrease	26	NC	Marg increase	Marg increase	Increase
(10) Diversification to NRE: *share of NRE in primary energy supply* (%)	0.22%	3.5%	NC	NC	NC	Marg increase	NC	NC	Marginal decrease

VII Resilience: *(11) Decentralization:* Number of biomass power producers	100	1,365	NC	Marg increase	NC	Marg increase		Marg decrease	Marginal decrease
VIII Environmental sustainability *(12) thousand tons of SO2*	1922 kt	43,179 kt (2016–25)	459 kt (2016–25)	$2x10^{-3}$ kt (2016–2025)	33 to 164kt (2016–25)	Marg inc	>260 kt (2011–25)	>45 kt to >242 kt (2016–25)	>797 kt to >1130 kt +2% to 3%
(13) Million tons of CO_2-equivalent	523	13,153 mt (2016–25)	180 mt (2016–25)	4 mt (2016–25)	10 to 49 mt (2016–25)	Marg inc	>78mt (2011–25)	>13 mt to >72 mt (2016–25)	>324 mt to >384 mt +3%

Source: Author

than double that in the dry season. Designers of the mainstream dams may have already anticipated[13] the impact of water resource developments listed in the MRC 20-year foreseeable future scenario (Table 4.3). However, the Thai irrigation expansion scenarios discussed in above (Case 2, +5 million ha) surpass the *total* basin irrigated area expansion in the MRC 20-year scenario (+0.8 million ha).

The range of hydrological futures studied by Hoanh et al. did not include levels of water diversions higher than those in the MRC 20-year scenario. This means higher-than-planned water diversions in Thailand may impact negatively on downstream hydropower, despite climate-induced increases in dry season flows (Table 4.3).

Perhaps because of combined uncertainty around upstream irrigation and climate change, Vietnam analysts regard any major upstream water diversions as challenging, notwithstanding the results of regional climate models projecting increased annual flows.

Provided they are able to negotiate with downstream interests, larger upstream farmers or agri-business interests emerge as winners of these natural resource development scenarios. This contributes to land consolidation and exit of smallholders. Such consolidation may generate local conflict, particularly if it is accompanied by perceptions of unfair manipulation of the land tenure system. Similarly, resistance may also develop among people who lose access to forests, fuelwood and other resources as a result of mining, rubber, or irrigation expansion. New grievances can catalyse contentious claim making in a context of increased political uncertainty (McAdam et al. 2001; Foran 2006). With Mekong mainstream hydropower development already controversial, a decision to proceed with the Sayaburi or other mainstream dam projects, as well as domestic coal and nuclear projects, is likely to feed advocates' ongoing demands for a re-evaluation of energy strategies. This could lengthening the already-long hydropower development life cycle, reducing financial returns to sponsors, but also encourage policy support for renewable energy and energy efficiency,

The six developments assessed however do not have a direct impact on access to electricity in low-access Cambodia and Myanmar: this is because increased access requires increased political commitment and capacity to expand access to rural users (World Bank 2011). Expanding access will be more challenging in a context of adaptation to climate change, where the needs of urban centres may dominate policy agendas. Equity impacts depend on governance responses and deserve more scenario analysis and distributional modelling (Warr et al. 2012; Fankhauser 2010).

6 Discussion and Conclusion

This chapter assessed how the Mekong region's energy system might respond to six sets of potential development decisions: mainstream hydropower dams, irrigation, sea-level rise, upland rubber, enhanced rail connectivity, and alumina refining and smelting. The analysis began by treating these as separate factors, then explored

[13] In the sense that their designed plant discharge (m^3/s) matches mean high season flow values under the MRC 20-year development scenario for the relevant reach of the river.

cumulative and emergent impacts on the energy system. The energy system was conceptualized as a multi-dimensional system whose key attributes include energy metabolism, governance, availability, access, affordability, long-term sustainability of supply, resilience, and environmental sustainability.

Several development projects assessed are uncertain with respect to scope, timing, and enabling conditions, and contingency is exacerbated by limited disclosure of investment plans. Our analysis responded using mixed quantitative and qualitative techniques structured by conservative assumptions. The scale and trajectories of the system are sufficiently large and entrenched that the selected development projects register as having minor impacts, including small but notable increases in energy demand (+3%), and in the case where mainstream hydropower is avoided, declining low-cost fossil reserves (−6%). The exceptional finding – that income poverty increases from energy price escalation due to sea level adaptation investments, as well as ethanol production – was difficult to quantify in light of uncertainties around the scope of adaptation impacts and governance responses, and remains important to explore further.

The analysis flagged likely of contention over increasing large power stations, and the equity impacts of adaptation responses to sea-level rise (Lebel et al. 2009). Contention is likely to increase as communities in a context of volatile and possibly lower average economic growth dispute distributional impacts of the infrastructural investments. Decisions to proceed with other developments might trigger unintended consequences on Mekong society through loss of access to traditional sources of energy.

The six sets of developments assessed here do not amount to investments in a more resilient or long-term sustainable energy system. However, the tools and techniques developed here can be used to explore the impacts of alternative investment strategies.

References

Abe, H. 2007. Potential for rural electrification based on biomass gasification in Cambodia. *Biomass and Bioenergy* 31: 656–664.

Agrawala, S., and S. Fankhauser. 2008. *Economic aspects of adaptation to climate change: Costs, benefits, and policy instruments*. Paris: OECD. Accessed 10 Feb 2013.

Asia Pacific Energy Research Center. 2009. APEC energy demand and supply outlook, 4th ed. Economy Review. http://www.ieej.or.jp/aperc/2009pdf/Outlook/Outlook_Volume_II_4E.pdf. Accessed 27 Aug 2010.

Asian Development Bank. 2005. Champasack road improvement project (Loan 1369-LAO[SF]) in the Lao People's Democratic Republic. Performance evaluation report. Manila: Asian Development Bank.

Asian Development Bank. 2009. *Building a sustainable energy future: The Greater Mekong Subregion*. 196 pages., Manila: Asian Development Bank.

Asian Development Bank. 2011. *Asian development outlook 2011. South-South economic links*. Manila: ADB.

Bonn2011. 2011. Messages from the Bonn2011 conference: The water, energy and food security nexus – Solutions for a green economy.

Brooks, D. H. 2008. Linking Asia's trade, logistics, and infrastructure. ADBI working paper series, no. 128. http://hdl.handle.net/10419/53651. Tokyo: Asian Development Bank Institute.

Canarail Consultants. 2011. Regional: Greater Mekong Subregion: Railway strategy study. Technical assistance consultant's report. Project number: 42518–01. June 2011. Manila: Asian Development Bank.

Carnegie Mellon University Green Design Institute. 2012. Economic input–output life cycle assessment (EIO-LCA) US 2002 (428) model [Internet]. http://www.eiolca.net/. Accessed 28 May 2012.

Chester, M.V., and A. Horvath. 2009. Environmental assessment of passenger transportation should include infrastructure and supply chains. *Environmental Research Letters* 4. doi: http://dx.doi.org/10.1088/1748-9326/4/2/024008

Chester, M.V., and A. Horvath. 2010. Life-cycle assessment of high-speed rail: The case of California. *Environmental Research Letters* 5: 014003.

EGAT. 2007. Thailand Power Development Plan (PDP 2007). Presentation by Prutichai Chonglertvanichkul. High level forum on Lao-Thai artnership in sustainable hydropower development. Shangri-La Hotel, Bangkok: Electricity Generating Authority of Thailand, 7 Sept 2007.

Fankhauser, S. 2010. The costs of adaptation. Wiley interdisciplinary reviews. *Climate Change* 1: 23–30. doi: 10.1002/wcc.14.

Federici, M., S. Ulgiati, and R. Basosia. 2008. A thermodynamic, environmental and material flow analysis of the Italian highway and railway transport systems. *Energy* 33: 760–775.

Floch, P., and D.J.H. Blake. 2010. Water transfer planning in Northeast Thailand: Rhetoric and practice. Working paper. Challenge program on water and food project No. 67. June 2010.

Foran, T. 2006. Rivers of contention: Pak Mun Dam, electricity planning, and state–society relations in Thailand, 1932–2004. Ph.D. Thesis, Division of Geography, Department of Geosciences, University of Sydney. http://hdl.handle.net/2123/1984. Accessed 23 Mar 2011

Foran, T., C.E. Greacen, and C. Middleton. 2008. Sustainable energy futures for the Mekong Region. Paper presented at Berlin conference on the human dimensions of global environmental challenge. http://web.fu-berlin.de/ffu/akumwelt/bc2008/download.htm. Accessed 23 Mar 2011

Foran, Tira. (forthcoming). Action research to improve Thailand's electricity planning processes. Chapter for M-POWER Book 4. In ed. R.C. Bastakoti. Vientiane: M-POWER Research Network. Accessed 23 Mar 2011

Foran, T., T. Wong, and S. Kelley. 2010. Mekong hydropower development: a review of governance and sustainability challenges. Final Draft., Norwegian water resources and energy directorate (NVE). http://www.mpowernet.org/download_pubdoc.php?doc=5056, Accessed 23 Mar 2011. Oslo.

Gerbens-Leenes, Winnie, and Arjen Y. Hoekstra. 2012. The water footprint of sweeteners and bio-ethanol. *Environment International* 40: 202–211. doi:10.1016/j.envint.2011.06.006.

Greacen, C.E., and A. Palettu. 2007. Electricity sector planning and hydropower. In *Democratizing water governance in the Mekong region*, ed. L. Lebel, J. Dore, R. Daniel, and Y.S. Koma, 93–126. Chiang Mai: Mekong Press.

Greacen, C.S., and C.E. Greacen. 2004. Thailand's electricity reforms: Privatization of benefits and socialization of costs and risks. *Pacific Affairs* 77: 517–541.

Harvey, L.D.D. 2010a. *Energy and the new reality*, Energy efficiency and the demand for energy services, vol. I. London: Earthscan.

Harvey, L.D.D. 2010b. *Energy and the new reality*, Carbon free energy supply, vol. II. London: Earthscan.

Hoanh, C.T., K. Jirayoot, G. Lacombe, and V. Srinetr. 2010. Impacts of climate change and development on Mekong flow regime. First assessment – 2009. MRC technical paper no. 29. Vientiane: Mekong River Commission.

ICEM. 2010a. MRC strategic environmental assessment of hydropower on the Mekong mainstream. Final report. Hanoi: International Centre for Environmental Management.

ICEM. 2010b. MRC strategic environmental assessment of hydropower on the Mekong mainstream. Impacts assessment (opportunities and risks) (Vol. Discussion draft. 14 May 2010). Hanoi: International Centre for Environmental Management.

IEA. 2009. World energy outlook: International energy agency. www.worldenergyoutlook.org/media/.../2009/weo2009.pdf, accessed 17th Sep 2012
IEA. 2011. *Energy balances of non-OECD countries 2011*. doi:10.1787/energy_bal_non-oecd-2011-en.
Integrietes Ressourcen Management (IRM-AG). 2008. Economics of energy integration: Application of MESSAGE model in the GMS. Draft final report. Vienna: Integrietes Ressourcen Management (IRM-AG).
Johnston, R., Chu Thai Hoanh, G. Lacombe, A. Noble, V. Smakhtin, D. Suhardiman, S.P. Kam, and P.S. Choo. 2010. Scoping study on natural resources and climate change in Southeast Asia with a focus on agriculture. Colombo: IWMI. http://publications.iwmi.org/pdf/h042414.pdf. Accessed 13 Oct 2010.
Kahrl, F., Y. Su, T. Tennigkeit, Y. Yang, and J. Xu. 2012. Large or small? Rethinking China's forest bioenergy policies. *Biomass and Bioenergy*. doi: 10.1016/j.biombioe.2012.01.042
Kamkongsak, L., and M. Law. 2001. Laying waste to the land: Thailand's Khong-Chi-Mun irrigation project. *Watershed* 6: 25–35.
Lacombe, G, S. Douangsavanh, B. Thepphavong, C.T. Hoanh, B. Bounvilay, A. Noble, O. Ongkeo, R. Johnston, and C. Phongpachith. 2012. *Is there enough water in the Vientiane plain? A water balance assessment of the Lower Nam Ngum basin*. IWMI.
Larsen, M. 2010. Lao PDR Development Report 2010. Natural resource management for sustainable development. Background paper. Economic assessment of the future of the lao mining sector. Washington, D.C: The World Bank.
Lazarus, K.M. 2009. *In search of aluminum: China's role in the Mekong Region*. Phnom Penh: Heinrich Böll Stiftung, WWF and International Institute for Sustainable Development.
Lebel, L., T. Foran, P. Garden, and J.B. Manuta. 2009. Adaptation to climate change and social justice: Challenges for flood and disaster management in Thailand. Ch. 9. In *Climate change adaptation in the water sector*, ed. F. Ludwig, P. Kabat, H. van Schaik, and M. van der Walk, 125–141. London: Earthscan.
Malik, U.S. 2009. Biofuels production for smallholder producers in the greater Mekong subregion. *Applied Energy* 86: S58–S68.
McAdam, D., S. Tarrow, and C. Tilly. 2001. *Dynamics of contention*. New York: Cambridge University Press.
Molle, F., and P. Floch. 2008a. The "Desert bloom" syndrome: Irrigation development, politics, and ideology in the Northeast of Thailand. In *Mekong program on water, environment and resilience*. Chiang Mai: Institut de Recherche pour le Développement, International Water Management Institute.
Molle, Francois, and Phillippe Floch. 2008b. Megaprojects and social and environmental changes: The case of the Thai "Water Grid". *Ambio* 37(3): 199–204.
Molle, F., P. Floch, B. Promphaking, and D.J.H. Blake. 2009a. The greening of *Isaan*: Politics, ideology, and irrigation development in Northeast Thailand. Chapter 10. In *Contested waterscapes in the Mekong Region: Hydropower, livelihoods and governance*, ed. F. Molle, T. Foran, and M. Käkönen, 253–282. London: Earthscan.
Molle, F., T. Foran, and M. Käkönen (eds.). 2009b. *Contested waterscapes in the Mekong Region: Hydropower, governance and livelihoods*. London: Earthscan.
Molle, Francois, Tira Foran, and Mira Käkönen (eds.). 2009c. *Contested waterscapes in the Mekong region: Hydropower, governance and livelihoods*. London: Earthscan.
Nakhooda, S., S. Dixit, and N.K. Dubash. 2007. *Empowering people. A governance analysis of electricity*. Washington, D.C: World Resources Institute.
Nguyen Anh Tuan. 2012. A case study on power sector restructuring in Vietnam. Pacific energy summit. 2012 summit papers.
Nguyen, T.L.T., S. Gheewala, and S. Garivait. 2007. Fossil energy savings and GHG mitigation potentials of ethanol as a gasoline substitute in Thailand. *Energy Policy* 35: 5195–5205.
Nguyen, T.L.T., S. Gheewala, and S. Garivait. 2008. Full chain energy analysis of fuel ethanol from cane molasses in Thailand. *Applied Energy* 85: 722–734.

Ribeiro, F.d M., and G.A. da Silva. 2010. Life-cycle inventory for hydroelectric generation: A Brazilian case study. *Journal of Cleaner Production* 18: 44–54. doi: 10.1016/j.jclepro.2009.09.006

Salvatore, M., and B. Damen. 2010. Bioenergy and food security. The BEFS analysis for Thailand. Environment and natural resources management working paper 42. Rome: FAO.

Sovacool, B.K. 2009. Reassessing energy security and the Trans-ASEAN natural gas pipeline network in Southeast Asia. *Pacific Affairs* 82: 467.

Sovacool, Benjamin K., Ishani Mukherjee, Ira Martina Drupady, and Anthony L. D'Agostino. 2011. Evaluating energy security performance from 1990 to 2010 for eighteen countries. *Energy* 36(10): 5846–5853. doi:10.1016/j.energy.2011.08.040.

Spath, P.L., M.K. Mann, and D.R. Kerr. 1999. Life cycle assessment of coal-fired power production. NREL/TP-570-25119. Golden: National Renewable Energy Laboratory.

Takeshita, T. 2009. A strategy for introducing modern bioenergy into developing Asia to avoid dangerous climate change. *Applied Energy* 86: S222–S232.

United Nations. 2010. Preview. Green growth, resources and resilience. Bangkok: UNESCAP, Environment and Development Division.

United Nations. 2012. World urbanization prospects: The 2011 revision: UN Department of Economic and Social Affairs (UNDESA), Population division.

Van Wee, B., P. Janse, and R. Van Den Brink. 2005. Comparing energy use and environmental performance of land transport modes. *Transport Reviews* 25: 3–24. doi: 10.1080/0144164100016 76861

Varun, R., and I.K. Bhat. 2010. Life cycle energy and GHG analysis of hydroelectric power development in India. *International Journal of Green Energy* 7: 361–375. doi: 10.1080/15435075.2010.493803.

Vivoda, V. 2010. Evaluating energy security in the Asia-Pacific region: A novel methodological approach. *Energy Policy* 38: 5258–5263. doi: ://data.worldbank.org/indicator

Warr, P., J. Menon, and A.A. Yusuf. 2012. Poverty impacts of natural resource revenues. *Journal of Asian Economics*. doi: 10.1016/j.asieco.2012.03.003

World Bank. 2011. *One goal, two paths. Achieving universal access to modern energy in East Asia and the Pacific*. Washington, D.C: The World Bank.

World Bank. 2012a. PovcalNet: An online poverty analysis tool [Internet]. http://iresearch.worldbank.org/PovcalNet/index.htm/%3E. Accessed 4 June 2012.

World Bank. 2012b. World development indicators. GDP 2011 in 2000 USD [Internet]. ://data.worldbank.org/indicator. Accessed 4 June 2012.

Ziegler, A.D., T.B. Bruun, M. Guardiola-Claramonte, T.W. Giambelluca, D. Lawrence, and T.L. Nguyen. 2009a. Environmental consequences of the demise in Swidden cultivation in Montane Mainland Southeast Asia: Hydrology and geomorphology. *Human Ecology: An Interdisciplinary Journal* 37: 361–373. doi: 10.1007/s10745-009-9258-x

Ziegler, A.D., J.M. Fox, and J. Xu. 2009b. The rubber juggernaut. *Science* 324: 1024–1025.

Chapter 5
Livelihoods and Migration

Lilao Bouapao

1 Methodological Background

Livelihood is a broad and wide ranging concept that considers the complex, multi-dimensional relationships between social and physical environments, and highlights the vulnerability context in which decisions about livelihood strategies occur. Exploration and elicitation of livelihood systems involves extensive contextual characterisation, comprised of endowments, ambient conditions and trends; livelihood-related resources; institutional processes; organizational structures; livelihood strategies; and livelihood outcomes (DFID 1999; Larson 2010). This discussion in this Chapter will rely on those indicators from the livelihoods strategies and outcomes framework that are salient for the assessment of envisaged future scenarios in the wider Mekong Region.

While livelihoods options are more likely to be observed through the lens of income, this assessment will expand the suite of metrics to include livelihood activities (occupation), migration, which indicates stability at the household level (Adger 1999), and consumption. A long list of relevant and useful indicators for livelihoods assessment can be identified. However variable data reliability and availability has limited the number of indicators that can be used to represent aspects of livelihood systems in the wider Mekong Region.

Migration is a critical resource dependency factor and depending on the type of migration can be classed as either a stabilising or destabilising livelihood component (Adger 1999). Migration is not necessarily a livelihood outcome; it is part of a livelihood strategy. In the Mekong delta a common livelihood diversification strategy is for the younger generation to migrate to urban centres to work in manufacturing or construction, although the livelihood of the parents remains reliant on subsistence agriculture (Pham 2010). Migration of some household members results in remittances to the household, enhancing household livelihood security and opportunities (Ellis 1998). Seasonal migration is also a prominent livelihood strategy to improve livelihood conditions. For example, farmers proximate to Cambodia's

A. Smajgl and J. Ward, *The Water-Food-Energy Nexus in the Mekong Region: Assessing Development Strategies Considering Cross-Sectoral and Transboundary Impacts*,
DOI 10.1007/978-1-4614-6120-3_5, © Springer Science+Business Media New York 2013

Tonle Sap travel long-distances once or twice a year to become seasonal fishers (Bouapao and Hall 2010). Migration of the whole household is often associated with coping with extreme events and generally occurs as a last resort (Davies 1996).

This Chapter reviews and synthesises literature based insights together with available primary and secondary data. Primary and census data were generated by the national statistical offices and the Mekong River Commission (MRC) Secretariat, in particular the Social Impact Monitoring and Vulnerability Assessment (SIMVA). Secondary data collated by the MRC and personal research activities are also considered. A case study approach is used to highlight specific livelihood aspects derived from the review, indicating a singular reliance on a one-size-fit-all method is likely to result in inconclusive livelihoods impact assessment. This assessment primarily draws on and develops existing multiple sectoral research, such as agriculture and fisheries, as these sectors are highly relevant for livelihoods in the Mekong Region. The structure of Mekong Region livelihood activities, migration patterns and influences, income distribution, and consumption are detailed and linked to the major influencing factors detailed in Chap. 1.

The following Section provides a status quo of livelihoods in the wider Mekong Region. Probable impacts of prospective changes on livelihoods will be provided in Section 3. Section 4 draws together all the analysis of livelihoods impacts of individual decisions, known as accumulative assessment. Conclusions are provided in the last Section.

2 Status Quo of Livelihoods

The majority of the wider Mekong Region population, estimated at 255 million, live in rural areas, are engaged in agriculture, and are dependent on natural resources for livelihoods, food, and income (MRC 2003, 2010; ADB 2007). Approximately 75 million of wider Mekong Region people belong to an estimated 200 ethnic groups, are typically concentrated in remote mountainous areas and are characteristically reliant on subsistence agriculture (ADB 2007). While economic and industrial development has contributed to reduce the direct dependence on agriculture for many households, rural livelihoods in the Mekong Region remain important to the majority of households (Thongbai 2009). ICEM (2010) concludes that despite the increasing rural to urban migration, driven by declining natural resources and growing employment opportunities in urban centres, the proportion of the rural compared to urban population is likely to remain relatively stable. The proportion of national rural populations in the wider Mekong Region ranges from 57% in Yunnan Province to 78% in Cambodia (Table 5.1). In 2000, the percentage of rural population in the wider Mekong Region countries was lowest in Myanmar (72%) and highest in Cambodia (84%) (The World Bank 2010). While the population in the wider Mekong Region are growing at rates ranging from 0.6% in Yunnan Province to 2.1% in Lao PDR (Table 5.1), resources relevant to livelihoods and food security are declining.

Table 5.1 Social and economic indicators in the wider Mekong region

Indicator	Cambodia	China Yunnan	Lao PDR	Myanmar (2009)	Thailand	Vietnam
Population 2007 (million)	14.4	44.8 (2005)	5.9	58.4	63.9	87.4
Mekong river basin population 2007 (million)	13.0		5.2		23.1	18.7
Average annual population growth rate (%, 2007)	1.9	0.6 (2005)	2.1	1.2	0.8	1.2
Rural population (% of total population, 2007)	78	57.0 (2005)	69	67.4	68	72
Population density (people per km2, 2007)	80		25		125	265
Composition of ethnic groups (% of total)	Khmer (90) Vietnamese (1) Mon-Khmer (2) Others (2)	Han 66% Tibeto-Burman 20% Sino-Thai 6% Hmong-Mien 3% Mon-Khmer 3% others 2%	Lao (55) Khmu (11) Hmong (8) Others (26)		Thai (75) Chinese (14) Others (11)	Kinh (Viet) (86.2) Tay (1.9) Tai (1.7) Muong (1.5) Khmer (1.4) Hmong (1) Others (4.1)
Access to clean drinking water (% of population, 2006)	65		60		98	92
GDP per capita (US$, 2006)	648		674		3000	900
Proportion of population below national poverty line (%)	35 (2004)	15 (2005)	33 (2003)	32.0 (2005)	10 (2006)	16 (2006)
Life expectancy at birth (years, 2007)	59	72.5 (2005)	64		71	74
Infant mortality (rate per 1000 live birth, 2006)	65	32	59		7	15
Malnutrition rate (%, 2007)	28		36 (2000)		7	20
Adult literacy rate (%, 2007)	76	90.9 (2005)	73	89.9 (2000)	94	90

Source: MRC 2010, China Yunnan province, UNEP and ADB (2007), Myanmar, ADB 2009b

Table 5.2 Employment in agricultural sector

	% of total labour force	Source
Cambodia	70.0	WHO 2009
China's Yunnan province	90.0	ADB and UNEP 2004
Lao PDR	78.5	Messerly et al. 2008; Sone 2009
Myanmar	50.0	ADB 2010
Thailand	39.4	National statistical office 2007
Vietnam	48.2	General statistics office 2010

2.1 Livelihood Activities

Although a shift is occurring from agricultural activities into other employment opportunities, employment of the labour force in agriculture range from more than 39.4% in Thailand to 90% in Yunnan Province, China (Table 5.2). In the lower Mekong Basin (LMB), approximately 70% of the total population who are 19 years and older, are farmers and fishers including farm and irregular labourers in the fresh water zone of the Mekong delta, Tonle Sap Lake, Siphandone, Udonthani, and Chiang Saen (Hall and Bouapao 2010). Note that the secrtoral employment estimate is for rural areas, excluding capital cities of provinces and districts.

However, many households in the Mekong Region depend on a combination of livelihood activities. A baseline survey on socioeconomics of households in eight fishing provinces in Cambodia undertaken during 1995–1996 shows that households were found to combine farming, fishing, fish selling, fish processing and different other activities for employment, income and food (Ahmed et al. 1998). Similarly, in the Songkram River Basin, Thailand, men and women are engaged in a range of occupations (Hortle and Suntornratana 2008).

In 2004, approximately 68.1% of Cambodia's rural employment was linked to agriculture, compared to 2.6% and 40.3% of employed persons in Phnom Penh and other urban areas, respectively (WFP 2010). Rural employment options are crop production (47.2%), livestock farming (13.6%), retail trade (10.3%), and manufacturing textiles, apparel and footwear (3.7%), while urban employment options are crop production (25.8%), retail trade (23.3%), livestock farming (7.2%) and fishing (6.5%). Data from the Basin Development Plan (BDP) (MRC 2010) show that close to 70,000 people are engaged in fisheries as their main occupation although the number is much higher if fishing is understood as a source of income and food. In the Tonle Sap, the number of people depending to varying degrees on fishing is frequently quoted to be 1.2 million people. A higher percentage of females compared to males work in agricultural employment (75.1% and 70.6% respectively). However, more males (2%) than females (0.9%) work in fisheries. Service and industrial employment account for 16% and 7.3% of the total employment, respectively.

In the Chinese Province of Yunnan agriculture is the single most important economic sector in the province. In addition to subsistence crops, rice, wheat, and other grains are produced as major cash crops. Although land suitable for cultivation is

Table 5.3 Percentage distribution of occupation in agriculture, fishery, and non-farming activities in the Lao PDR

	Total active population	Farmer	Fisherman	Livestock farmer mainly	Mixed farmer	Non-farm activity	Total
Total	2,738,892	64.3	0.1	0.2	14.0	21.5	100
Female	1,375,450	66.3	0.0	0.2	14.8	18.7	100
Male	1,363,442	62.2	0.1	0.2	13.2	24.3	100

Source: Department of statistics, Lao PDR

Table 5.4 Number and percentage of employed persons in Thailand by industry: 2001–2007

	2001	2002	2003	2004	2005	2006	2007
Total ('000)	32,104.2	33,060.9	33,841.0	34,728.8	35,257.2	35,685.5	36,249.5
Agricultural ('000)	13,611.8	14,041.8	13,880.1	13,633.9	13,617.0	14,170.5	14,306.0
Non-agricultural ('000)	18,492.4	19,019.0	19,960.9	21,094.9	21,640.2	21,515.0	21,943.4
Agricultural, %	42.4	42.5	41.0	39.3	38.6	39.7	39.4
Non-agricultural, %	57.6	57.5	59.0	60.7	61.4	60.3	60.6

Source: National statistical office 2007

limited, Yunnan farmers make up of 80–95% of the labour force, with intensive cultivation.

In 2005, 64% of the total working population of Lao PDR were engaged as farmers, 0.2% as livestock farmers, 14% as mixed farmers. Non-farming households account for 21.5%. Fishing as an occupation represented less than 1% of the workforce. However, almost all households are engaged in fishing to some degree. More females than males are engaged in farming activities (66.3% and 62.2% respectively), while more males than females are engaged in non-farm activities (24.3% and 18.7% respectively) (Table 5.3).

Myanmar's main economic activity is highly agro-based, including fisheries and forestry, contributing approximately 40% to GDP, and employing more than 50% of total labour force (ADB 2009a). In Thailand, the proportion of population involved in agricultural activities is less than those in non agricultural employment and the trend is declining (Table 5.4). In 2007, 39.4% of the total employed persons were involved in agriculture and 60.6% were engaged in non-agricultural activities. While employment grew in all sectors, the share of employment in the agricultural sector declined by 3% over the period 2001–2007. However, aggregate decline disguises the variation in sectoral employment across regions in Thailand. The SIMVA data (MRC 2010) indicate that the proportion of rural population engaged in farming, and farming related activities such as farm and irregular labour in Chiang Saen and Udonthani, account for 71% of the total households. Permanent employment and businesses account for 7% each.

The sectoral and spatial pattern of occupational change in observed in Thailand is similar to that observed in the Vietnamese Mekong Delta for the period of 2004–2008. The proportion of the population engaged in farm wage and farm self employment declined, while the non-farm wage and non-farm self-employment increased

Table 5.5 Primary occupation of workforce aged 15 years and over in the Mekong delta, Vietnam; 2004 and 2008

	Total	Non-farm wage employment	Farm wage employment	Farm self employment	Non-farm self employment
2004	100	20.1	13.2	46.1	20.6
2008	100	23.6	10.4	44.7	21.3

Source: GSO 2010

(Table 5.5). Data from GSO (2010) show that while the proportion of aged 15 years and older engaged in agriculture, forestry, and fishery declined by more than 4% from 2004 to 2008, there is no significant increase in any particular other sector, implying a major shift from agriculture, forestry, and fishery to other sectors, and the redistribution is quite even.

The proportion of the population engaged in farm wage and farm self-employment is significantly higher in rural areas than urban areas in 2004 (66.5% and 16.2% respectively). The proportion of farm wage and self employment decreased in 2008 for both rural and urban areas (GSO 2010).

2.2 *Income Sources*

SIMVA data (MRC 2010) indicates that rice and remittances are the primary sources of income in the LMB, contributing income of 50% and 30% of the households, respectively. The sale of fish, irregular/seasonal, and full-time employment, each contributes to the income of 25% of the total households. In the Tonle Sap Lake area (Cambodia) and the Siphandone area (Lao PDR), fish sales generate income for approximately 40% of households. In Cambodia, a large proportion of people depend on forestry and fisheries as their sources of income, especially in times of hardship. Approximately 16% of the poor derive more than 50% of their income from forestry and fisheries (WFP 2010).

Results of the Lao Expenditure and Consumption Survey (LECS 4) shows that rice represents a high proportion of total household income in all provinces, except for households in Houaphanh and Vientiane provinces, who obtain their income mainly from vegetables and fruits (Department of Statistics 2009). Red meat, fish and vegetables account for more than 83% the food items consumed in 2007/08. There is an increasing trend of rice consumption. The proportion of rice consumed increased by more than 3% during 2002/2003 to 2007/2008, while the proportion of red meat and fish dropped slightly over the same period. This is consistent with data from other sources, showing a decline in the production of both fish capture and culture over the last few years (MOFA 2007, 2008). Nevertheless, a move away from natural resources dependency to livestock husbandry has already taken place. Data of LECS 2002/2003 and 2007/2008 show that ownership of fishing nets and boats decreased, while the average number of cattle per household increased. By ethnicity, the Hmong-Mien groups report the highest involvement in livestock husbandry, followed by the Lao-Tai groups (WFP 2007).

Table 5.6 Structure of monthly income per capita by sources of income, %, in the Mekong Delta

	Salary or wage	Agriculture	Forestry	Fishery	Industry	Construction	Trade	Service	Others
2004	25.7	28.6	0.4	9.9	3.3	0.7	10.6	6.8	13.9
2008	26.0	29.9	0.3	8.8	3.6	0.3	10.1	6.7	14.3

Source: GSO 2010

Table 5.7 Most important occupations for the household by country study sites

Most Important Occupation in Household	Tonle Sap Lake, Cambodia	Siphandone, Laos	Chiangrai and Udonthani	Mekong Delta	Total
Farming	61.80%	87.60%	77.60%	63.80%	72.70%
Fishing	25.30%	3.20%	0.90%	2.60%	8.00%
Collecting OAAs	0.00%	0.00%	0.00%	0.30%	0.10%
Collecting useful plants	0.00%	0.00%	0.30%	0.30%	0.10%
Other fish and water related occupations	1.80%	0.00%	2.10%	0.60%	1.10%
Farm labour	1.80%	0.00%	0.00%	9.70%	2.90%
Other irregular work	0.90%	0.30%	6.50%	3.20%	2.70%
Permanent employment	0.90%	3.50%	3.50%	10.00%	4.50%
Business	3.20%	3.20%	7.40%	4.40%	4.60%
Other	3.20%	2.10%	1.50%	5.00%	2.90%
Total	100.00%	100.00%	100.00%	100.00%	100.00%

Source: Hall and Bouapao 2010

In the Mekong delta, income from natural resources, particularly from forest and fishery declined between 2004 and 2008. During the same period income from salaries or wages and from sales of agricultural produce increased slightly (Table 5.6). The contribution of income from fishery in the Mekong delta declined from 9.9% in 2004 to 8.9% in 2008 (GSO 2010).

The following Section reports specific components of the SIMVA data, elicited in 2009 (MRC 2010), as a baseline for comparisons of changes to livelihoods water as result of impending wider Mekong Region decisions on water management, energy, mining, land use and transport infrastructure interactions forecast to occur in future scenarios.

Farming is the most important livelihoods activity for the majority of people in Tonle Sap, Siphandone, Chiangrai and Chiang Saen and Mekong delta (Table 5.7).

Figure 5.1 illustrates that the most common sources of income are from the sale of rice (50%), followed by remittances (31%), irregular/seasonal employment (30%), full-time employment (25%), sale of livestock (25%), sale of fish catch (25%), and business profit (19%). In the Tonle Sap Lake and Siphandone, fish sales represent the primary income of more than 40% of households, compared to 10% in

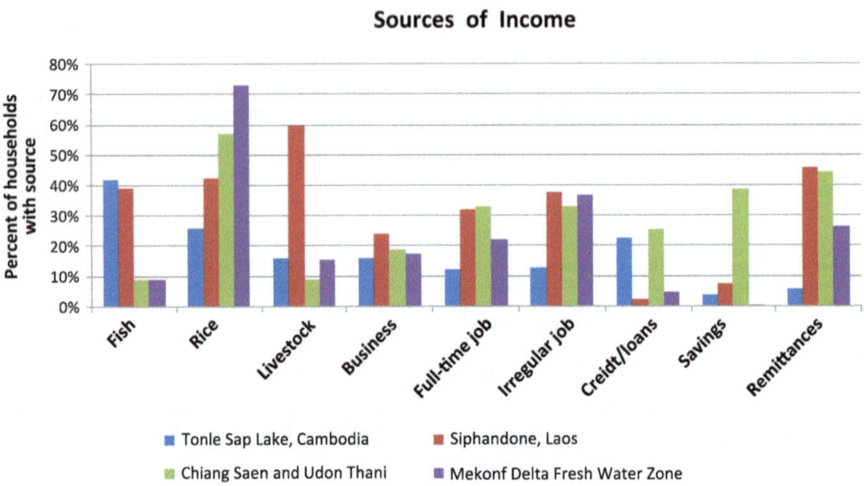

Fig. 5.1 Sources of income in wider Mekong region countries (Source: Hall and Bouapao 2010)

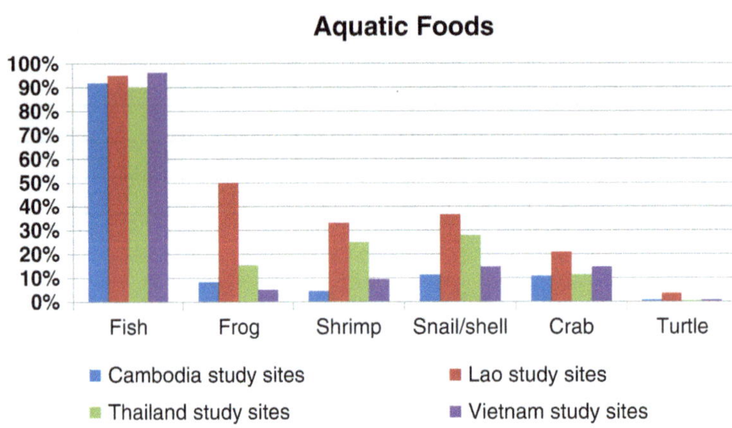

Fig. 5.2 Aquatic food (Source: Hall and Bouapao 2010)

Thailand and the Mekong delta. In the Mekong delta, rice is a source of income for more than 70% of households, while rice accounts as a major source of income for 55% in the Thai study areas.

Rice and other crops are likely to remain as the major staple foods in the Mekong River Basin for the next decades (Pech and Sunada 2008). In 2009, the SIMVA results indicate that the basic types of food eaten by rural households were rice, fish, and vegetables (99%, 93%, and 95% respectively). There is considerable diversity in the other foods and sources of protein (Fig. 5.2).

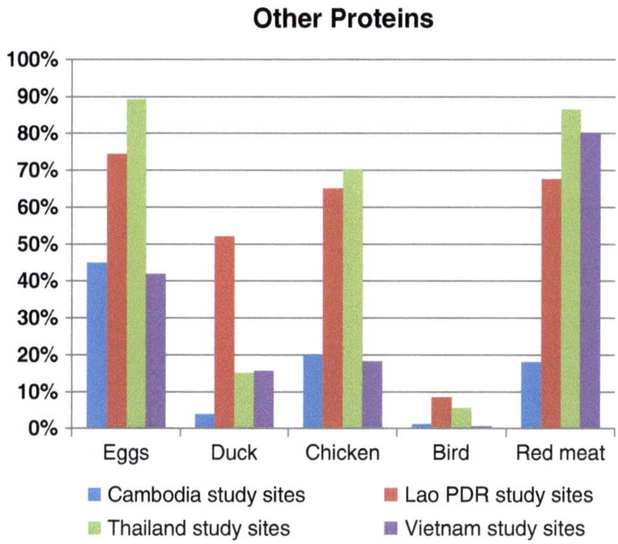

Fig. 5.3 None aquatic food (Source: Hall and Bouapao 2010)

Similar to the sources of income, households in the Tonle Sap area rely heavily on fish as a primary food source. Chicken and red meat are sources of income for very small proportion of households in Cambodia. In Chiang Saen and Udonthani, fish, eggs, chicken and red meat are the main sources of food for the majority of the households. In the Mekong delta, key sources of food are fish, eggs, and red meat.

The mean daily calorie intake in the Tonle Sap area, in Siphanhdone, in Chiang Saen, Udonthani, and in the Mekong delta is 2,407 cal, which is above the regional recommended minimum of 2,100.[1] The majority of calories are derived from rice (76%), partly because of its status as a staple food, and because of its high calorific value (3,550 cal/kg). Aquatic resources contribute 13% and other foods 11%.

According to Fig. 5.4, most household food consumption in rural Lao PDR in the Mekong corridor is natural or self produced, whilst most food consumed by rural people in Thailand, Cambodia's Tonle Sap Lake, and Vietnam's Mekong corridors is purchased.

Overall *cash income* in these areas is derived largely from sources, which are not directly affected by changes in the river flow and resource availability. The exceptions are the Tonle Sap, where over one third of the total income is from fish, and the Mekong delta, where one quarter of income is from crops (Fig. 5.5). The primary income sources of households located in Thai mainstream areas are derived largely from non-aquatic resources.

It should be noted that fishing households around the Tonle Sap derived over two thirds of their household income from fishing, far more than those in other areas,

[1]FAO standard, as supplied by the National Statistics Office of Lao PDR.

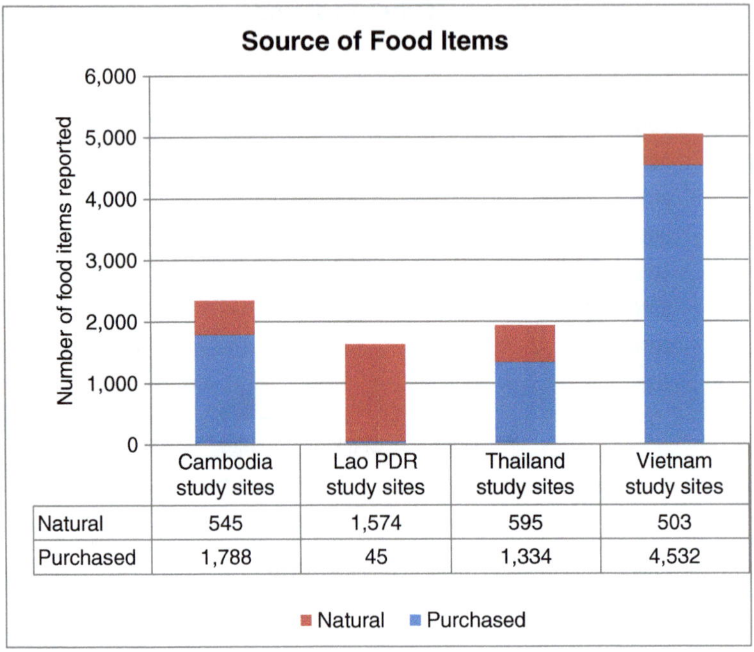

Fig. 5.4 Sources of food items (Source: Hall and Bouapao 2010)

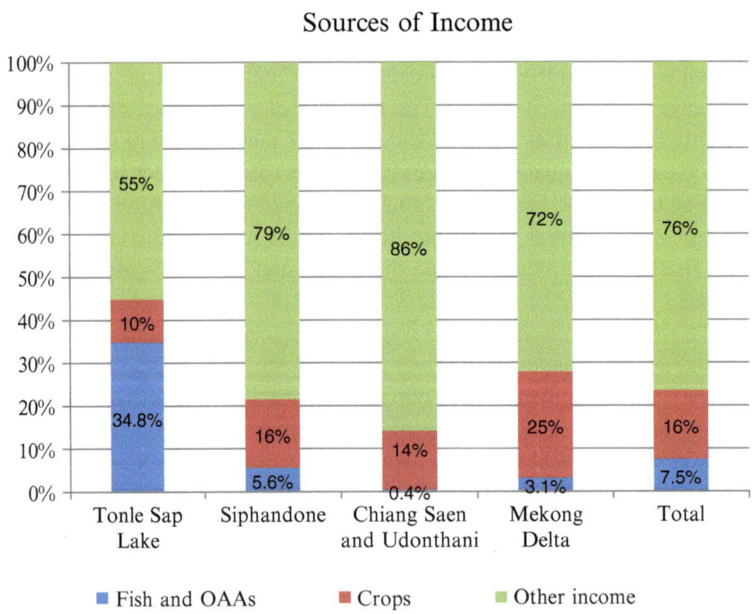

Fig. 5.5 Share of income (Source: Hall and Bouapao 2010)

Table 5.8 Incidence of urban and rural poverty by country

Country	Urban	Rural	Ratio
Cambodia	18.2	40.1	2.20
Lao PDR	26.9	41.0	1.52
Thailand	4.0	12.6	3.15
Viet Nam	6.6	35.6	5.39

Source: UNESCAP (undated)

where alternative sources of income are accessible. Those households who use paddies, ponds, and/or canals as preferred fishing resources obtain the highest percentage of income from crops (34%).

2.3 Livelihoods Outcomes

The last two decades have seen rapid economic growth in the wider Mekong Region. However, the quality of life of the poorest has only marginally improved. The majority of the rural population continue to live in great poverty (BDP 2006). The proportion of the population living below the national poverty line (Table 5.1) is highest in Cambodia (35% in 2004), followed by Lao PDR (33% in 2003), and Myanmar (32% in 2005). In Cambodia, 35% of the population live below rural and urban poverty lines of US$0.46 and US$0.63 (WHO 2009). The data reported in Table 5.8 illustrates that the degree of poverty incidence in urban areas is much lower than rural areas in all countries. The gap of incidence of poverty[2] between urban and rural areas is largest in Vietnam (more than five times), followed by Thailand (more than three times), Cambodia (more than two times), and Lao PDR (1.5 times). Countries with the lowest incidence of poverty have the biggest difference between urban and rural areas.

Overall, poverty rates are higher in communities located in remote upland areas and lower when located closer to lowland areas. However, due to the high population densities in the latter, the absolute number of poor is higher in lowland areas closer to the mainstream Mekong (ICEM 2010). Prevalence of under-weight children under five of age is also highest in Lao PDR (37%). Cambodia and Lao PDR have the most acute development indicators such as GDP per capita, life expectancy at birth, infant mortality, malnutrition, and adult literacy rates. Cambodia has improved access to clean drinking water, malnutrition, and adult literacy rates compared to Lao PDR, but diminished development when measured by GDP per capita, life expectancy, and infant mortality rates Table 5.1. Overall, educational levels in the countries of the LMB are low. The average school attendance is less than 5 years. Population growth

[2] The incidence of poverty refers to the proportion of households with per capita incomes below the national poverty threshold.

remains high despite the decline in fertility rates and improved health conditions and life expectancy. Among the wider Mekong Region countries, population growth rate is highest in the Lao PDR (2.1%), followed by Cambodia (1.9%).

2.4 *Migration*

A number of distinct and established patterns of migration in the wider Mekong Region involve the migration of:

1. Predominately unskilled labour from Cambodia, Lao PDR, and Myanmar into Thailand,
2. Higher skilled workers from Vietnam into Cambodia and Lao PDR, and
3. High skilled people from Yunnan Province into Lao PDR and Myanmar (Revenga et al. 2006)

Most of the outward migration is from Cambodia, Lao PDR, and Myanmar, and most of the inward migration is into Thailand, with approximately two million migrants (Lewis et al. 2010; Mekong Migration Network and Asian Migrant Centre 2009). Migrants from Myanmar accounted for the majority of migrants in Thailand and the migrants grew significantly during 2001–2004 (Revenga et al. 2006). Note that it is not clear, however, that the increase in the number of migrants in Thailand during that period was a result of improvement in the registration or is actually an increase in the migration flows (ibid). The proportion of women from Cambodia, Lao PDR, and Myanmar migrating to Thailand grew from approximately 30% in 1998 to 45% in 2005, with Lao migrant women outnumbering migrant men (Caouette et al. Undated).

The Lao Population Census 2005 indicates that approximately 500 people moved into Lao PDR, while 8,500 people left Lao PDR for other countries (Steering Committee for Census of Population and Housing Department of Statistics 2006). In contrast to a previous history as primarily an emigration province, the direction of migration in Yunnan Province has become a more bi-directional process (Mekong Migration Network and Asian Migrant Centre 2009). Migrants in Thailand are employed in sectors requiring unskilled labour, satisfying the excess labour requirements that remained unfulfilled from the local Thai work force. Examples include cleaning, fisheries, construction, and other dirty, dangerous, and difficult work (Lewis et al. (2010).

There is an increasing trend in labour migration from more advanced economies to poorer countries in the wider Mekong Region. While, Lao PDR is predominantly a source country for migrants to Thailand, it is also a target country for migrants from neighbouring countries, particularly from Vietnam and China. The number of migrants from China is increasing rapidly because of the number of development projects (Mekong Migration Network and Asian Migrant Centre 2009). Chinese workers are increasingly moving to Cambodia to work often in Chinese businesses. In some cases such as in Thailand, industries establish themselves at border locations to take advantage of abundant and low-cost labour in neighbouring countries

(Caouette et al., undated). There are large numbers of semi-skilled temporary migrants from China and Vietnam who remain committed to national enterprises established in other wider Mekong Region countries. It is estimated that 40,000 Chinese people are working in Myanmar for Chinese companies, and 80,000 Chinese working in Lao PDR for Chinese led projects (Lewis et al. 2010).

There is also a substantial daily migration across borders to work in unskilled jobs including agriculture and trade (Lewis et al. 2010). Note that movement across borders has long been a feature of daily life of ethnic groups living along the borders of the wider Mekong Region countries, who often have cross border kin-relationships with people from the same ethnic groups (Mekong Migration Network and Asian Migrant Centre 2009).

3 Assessing Impacts of Prospective Changes on Livelihood and Migration

3.1 Mainstream Dams

The Nations (2010) argues that up to 40 million people could be affected if the mainstream dams on the Mekong proceed. Osborne (2010) suggests the livelihoods of 29.6 million people in Lao PDR, Thailand, and Cambodia and 14 million people in Vietnam could be negatively affected. Assuming mainstream dams cause decline in captured fish and the decline is distributed evenly throughout the LMB corridor, results of the SIMVA survey suggests about 93% of the households stated they consumed fish in the preceding week would be impacted. Considering that 25 million people live in the 15 km corridor of the Mekong it documents a high dependency on the availability of fish. Although the dependency on fish increases with the proximity to the river, the loss of fish impacts on households well beyond the 15 km corridor (Bouapao and Hall 2010). About 80% of the fish sold in the local markets of Champasak are from the Khong District of the Mekong (Baird et al. 2001). Twenty-five percent of the total fish catch of the Tonle Sap Lake is consumed by local fishing households, indicating that most of the catch is for consumers located in centres beyond the lake boundaries such as Phnom Penh (Hall and Bouapao 2010).

According to the national consultation workshop on the proposed Xayaburi hydropower dam on the Mekong mainstream, held in Can Tho on 14 January:

> The workshop participants expressed their great concerns and worries about the possible impacts of the project on the productivity of the Mekong delta and the livelihood of millions of people living in the delta. Potentially acing [sic] double impacts, from climate change and sea water level rise, and from the fast development in the Mekong upstream part, especially on the mainstream, the delta would be adversely threatened by severe impacts resulting in the intrusion of saline sea water far into the inland, immense damage to the fisheries, declination of agro-productivity vital to millions and unpredictable degradation of the invaluable biodiversity, cited most of the comments at the workshop (Vietnam National Mekong Committee, undated).

Table 5.9 Impacts of mainstream dams

Indicator	Occupation	Income	Food	Migration
Cambodia	No change	Decrease	Decrease	Increase migration into Thailand and Vietnam
China's Yunnan Province	No direct trans-boundary impacts	No direct trans-boundary impacts	No direct trans-boundary impacts	No direct trans-boundary impacts
Lao PDR	No change	Decrease	Decrease	Increase migration into Thailand and Vietnam
Myanmar	No change	n/a	n/a	n/a
Thailand	No change	Negligible	Negligible	Negligible
Vietnam	No change	Decrease	Decrease	Increase migration from Cambodia and Lao PDR

Note: n/a – not available

3.1.1 Foreseeable Future Scenario

Based on outputs for the biophysical impacts of the construction of all 12 mainstream dams estimated by the Basin Development Plan Programme, Phase 2 (2010), Podger et al. (2004) and ICEM (2010), the livelihoods impacts are summarized in Table 5.9 and are subsequently discussed.

3.1.2 Occupation

Based on the estimated loss of fish reported in the Basin Development Plan Programme, Phase 2 (2010), if all dams are built, the overall loss to fish catch in the river and its floodplain is estimated to be 58% of the baseline yield of 593,000 t/year. By countries, the decline is 84%, 63%, 41%, and 40% for Lao PDR, Cambodia, Thailand, and Vietnam, respectively. This alone does not indicate that fishing households would give up fishing. The SIMVA results show that 37% of fishing households would give up fishing if the daily catch was less than 1 kg. Estimates of fish catch based on the SIMVA data indicate that the decline in all countries does not reduce the catch below 1 kg/day. This implies that the number of fishing households in Cambodia, Lao PDR, Thailand, and Vietnam would remain stable.

However, by 2030 the proportion of fishing households would decline due to the combination of population growth and the decline in fish. Observations made in the context of tributary dams such as Pak Mun indicate a decline in fish catch of 50–100% resulting in a drop of fishing households in areas upstream of the dam ranging from 66.7% to 95.6% (Amornsakchai et al. 2000).

Table 5.10 Calorie intake by study sites

Study Sites	Fish calorie intake 2009	Total calorie intake 2009	Decline in fish due to 12 dams	Impacts
	Calories	Calories	%	Calories
Tonle Sap Lake, Cambodia	335	2,121	63	1,910
Siphandone, Lao PDR	300	3,171	84	2,919
Chiang Saen and Udonthani, Thailand	281	2,471	41	2,356
Mekong Delta, Vietnam	301	1,864	40	1,744
Total	**303**	**2,407**	58	2,231

Source: Hall and Bouapao 2010

3.1.3 Food

The estimate loss of fish due to the predicted dam construction and operation will result in less food. The average calorie intake in the Tonle Sap area in 2009 was 2,121 cal per capita per day. Fish contribute 335 cal or 16% of the total calorie intake. A 63% reduction of the fish catch will reduce the per capita per day calories to 1,910 (Table 5.10). This is less than the regional recommended minimum of 2,100 cal.[3] Moreover, the loss of fish is significant well beyond its calorific values. Fish contain essential micro nutrients not found in rice (or other staple foods), as well as fatty acids that are essential for the development of the brain and body. Micro nutrients are recognized as highly important, particularly in the diets of pregnant women, infants and young children (FAO 2005). Similarly, other sites will be affected, including the Mekong delta, where the overall calorie intake was 1,864 cal per capita per day.

3.1.4 Income

A 63% loss of fish capture will have considerable impacts on the income of people in the Tonle Sap area. The SIMVA results show that the average per capita income in the Tonle Sap area was just slightly above US$1 per day, considered by the World Bank as the minimum required for daily subsistence in a developing country. US$2 per day is recommended as a more realistic number for development planning.[4]

[3] FAO standard, as supplied by the National Statistics Office of Lao PDR.

[4] World Bank (2004), Millennium Development Goals, http://web.worldbank.org/WBSITE/EXTERNAL/EXTABOUTUS/0,,contentMDK:20104132~menuPK:250991~pagePK:43912~piPK:44037~theSitePK:29708,00.html. Accessed on 25 August 2010.

Table 5.11 Income per capita per day

Study Sites	Per capita income per day	Fish contribution	Decline in fish due to 12 dams	Impacts
	$US	$US	%	$US
Tonle Sap Lake, Cambodia	1.02	0.36	63.0	0.80
Siphandone, Lao PDR	1.43	0.08	84.0	1.36
Chiang Saen and Udonthani, Thailand	3.48	0.01	41.0	3.47
Mekong Delta, Vietnam	1.57	0.05	40.0	1.55

Source: Hall and Bouapao 2010

Income from fish and other aquatic animals (OAAs) contributes 35% to household income. A 63% loss of fish results in a household income of less than 1 US$ (Table 5.11). In Siphandone and Mekong delta, the decline in fish will force household income in these sites closer to the poverty line. In the Thai sites, however, the share of income from fish in the total income is small, hence there is likely to be very small impacts on income.

The estimated losses of agricultural lands, river bank gardens and biodiversity will further deteriorate the food availability and reduce income. With the estimated losses in these resources, it is likely that there will be increased reliance on OAAs, which have been traditionally used during the dry seasons and drought years, when less fish are available (Meusch et al. 2003; Mollot et al. 2003; Balzer et al. 2005; Halwart 2008).

However, in the Mekong delta, the increase in dry season flow possibly caused by the mainstream dams, will reduce salinity, hence increasing rice yield. This could lead to an increase in income from rice.

3.1.5 Migration

Losses in access to fish, riverbank gardens and inundated agricultural areas will be important push factors for migration from rural to urban areas and from Cambodia and Lao PDR to Thailand. Observations from cases such as the Pak Mun dam (Amornsakchai et al. 2000) show that those members of households affected by the loss of agricultural lands, river-bank garden, and fisheries are likely to migrate to either urban areas or to forest reserve areas (or other common properties), as compensation may not be enough to buy alternate land. These groups will become more acutely vulnerable because of the partial sale of their livelihoods assets due to the shift from farm based occupation and decline in grazing lands. It is also

Table 5.12 Impacts of water diversion

Indicator	Occupation	Income	Food	Migration
Cambodia	Negligible	Decrease income from fisheries, river bank garden and farming	Decrease food from fisheries	Increase migration into Thailand
China's Yunnan province	No direct trans-boundary impacts	No direct trans-boundary impacts	No direct trans-boundary impacts	No direct trans-boundary impacts
Central Lao PDR	Negligible	Decrease income from fisheries, riverbank garden and farming	Decrease food from fisheries	Increase labour migration into Thailand
Myanmar	No direct trans-boundary impacts	No direct trans-boundary impacts	No direct trans-boundary impacts	No direct trans-boundary impacts
Northeast Thailand	Increase in commercial agriculture. Reduced subsistence farming	Increase income from wage	Increase purchased food	Increase labour migration from Cambodia and Lao PDR
Vietnam	Negligible	Decrease income from fisheries and farming	Decrease food from fisheries	Negligible

Note: n/a: not available

likely that by 2030, people in Cambodia and Lao PDR will migrate to Vietnam for wage employment.

Very few local people in Cambodia and Lao PDR, where the majority of the 12 dams would be constructed, would be qualified for the construction jobs. A large proportion of the dam workforce therefore, would come from other countries, particularly dominated by Chinese construction contractors.

3.2 Lao PDR Water Diversion; Specifically to North East Thailand

Lessons on the impacts of water diversion on livelihoods have been well documented. Examples include the case of water diversion from the Nam Song River into Nam Ngum one reservoir. The diversions caused the loss of agricultural lands, river-bank gardens, fisheries, lack of water for drinking and washing affecting more than 1,000 households residing along the Nam Song River, and other property such as boats and fishing nets. Plausible impacts caused by the water diversion are summarized in Table 5.12.

Table 5.13 Percentage of average monthly household income in northeast Thailand by source of income

	1975-76	1981	1986	1988	1990	1992	1994	1996	1998	1999*	2000	2001**
(Value : Baht)	1,928	3,378	3,631	4,106	5,625	7,062	8,262	10,779	12,492	12,729	12,150	12,185
Total Income	100	100	100	100	100	100	100	100	100	100	100	100
Wages and Salaries	28.3	26.7	33.7	34.4	36.3	39	41.2	39.9	40.2	41.1	42.4	44
Profits, Non-farm	20.6	17.1	17.6	15.7	17.3	18.5	19.1	19.9	18.5	19.2	18.5	17.1
Profits from Farming	19.3	20.8	15.4	15.9	15.7	12.6	11.1	11.9	11.2	9.4	9.6	9.4
Property Income	0.8	0.9	0.9	0.9	1.1	1.5	1.1	1.6	1.9	1.9	1.5	1.8
Current Transfers	3.9	5.2	5.6	7	5.6	5.8	7.1	7.2	7.9	8.6	8.3	8.7
Income in-kind	26	27.8	25.8	24.8	21.2	21.1	19.3	18	18.5	18.3	18.4	17.8
Other Money Receipts1	1.1	1.5	1	1.3	2.8	1.5	1.1	1.5	1.8	1.5	1.3	1.2

Source: National statistical office, Thailand

[a]Data collection period was from June – September 1999

[b] Data collection period was from March – May, August – October 2001

Table 5.14 Monthly household income in northeast Thailand by source of income

	1975-76	1981	1986	1988	1990	1992	1994	1996	1998	1999*	2000	2001**
(Value : Baht)	1,928	3,378	3,631	4,106	5,625	7,062	8,262	10,779	12,492	12,729	12,150	12,185
Total Income	100	100	100	100	100	100	100	100	100	100	100	100
Wages and Salaries	546	902	1224	1412	2042	2754	3404	4301	5022	5232	5152	5361
Profits, Non-farm	397	578	639	645	973	1306	1578	2145	2311	2444	2248	2084
Profits from Farming	372	703	559	653	883	890	917	1283	1399	1197	1166	1145
Property Income	15	30	33	37	62	106	91	172	237	242	182	219
Current Transfers	75	176	203	287	315	410	587	776	987	1095	1008	1060
Income in-kind	501	939	937	1018	1193	1490	1595	1940	2311	2329	2236	2169
Other Money Receipts1	21	51	36	53	158	106	91	162	225	191	158	146

[a]Data collection period was from June – September 1999

[b]Data collection period was from March – May, August – October 2001

In northeast Thailand, increased water availability due to proposed water diversion is likely to contribute to increasing mechanization and consolidation of land holdings and an expansion of commercial agriculture (ICEM 2010). But the expansion of large scale-commercial agriculture means reduction in subsistence agriculture (or small farmers), reinforcing a current trend. Thai national statistics indicates that the percentage of population engaged in agriculture declined, while the non-agricultural sector increased over the period 2001–2007. Consistent with national trends, the percentage of income from wages and salaries, both in structure (Table 5.13) and absolute value (Table 5.14) increased in northeast Thailand.

Reduced dry-season flow will have negative impacts on local and downstream fisheries (Johnston et al. 2009) and agriculture. It is likely that the proposed 300 m^3/s water transfer from the mainstream Mekong (assuming the outtake points are around Vientiane and Bolikhamxay province), will have the most impact on 2.3 million households in the 15 km corridor of the Mekong delta (Table 5.15). Changes to these households engaged in river bank garden and farming, is most likely due to the dry season salinity intrusion into agricultural areas, and subsequent reductions in productivity. The lower level of the flow of the Mekong during the dry season might have impacts on more than 1.2 million households involved in river-bank garden, farming, and fishing in the Cambodia corridor and about 151,000 households

Table 5.15 Employment by farming, fishing, and non-farming sectors in the lower Mekong basin countries

	Cambodia	South of Laos	Vietnam	Total
15km from either side of the Mekong	9,895,523	740,809	14,000,000	24,636,332
% LMB Pop	87.10%	82.00%	85.80%	
Rural population	8,619,000	592,647	12,000,000	21,211,647
Average household size	6.1	4	5.1	
Rural Households	1,412,951	148,162	2,330,328	3,891,441
% of river garden households	7.00%	13.80%	29.40%	
River Garden Households in rural areas	98,907	20,446	685,116	804,469
% of farming households	60.00%	85.00%	65.00%	
Farming households	847,770	125,937	1,514,713	2,488,421
% of fishing households	20.00%	3.00%	2.90%	
Fishing households	282,590	4,445	67,580	354,615
% of non-farm households	20.00%	12.00%	32.10%	
Non-farm households	282,590	17,779	748,035	1,048,405
Total farm and non-farm households				18,482,268

Source: Estimate based on and adjusted from various sources including SIMVA, MRC database, general statistics of Vietnam and department of statistics of Lao PDR. Population within 15 km corridor is derived from Lanscan 2007

engaged in water related activities in the southern part of Lao PDR (including Khammuane, Savannakhet, and Champasak provinces).

Based on the fact that decline in fishing alone does not indicate that fishing households would give up fishing (discussed earlier in this paper) and based on knowledge and experience of the author, occupational changes possibly caused by the water diversion would be negligible. However, income and food of those people along the Mekong corridor would decrease unless alternate sources are accessible. This contributes to the push factor for increased migration from Cambodia and southern Lao PDR into Thailand. As a result it is likely that the proportion of income from remittances will increase in the Cambodia and southern Lao PDR.

3.3 Sea-Level Rise

Livelihood impacts of climate change-induced sea-level rise are mediated through reductions in rice productivity, rather than direct effects on towns and villages (Le et al. 2007). Wassmann et al. (2004) argue that reduced rice production will occur due to increased flooding in tidally inundated areas and prolonged flooding periods. In the Mekong delta, under the sea level rise scenario of 20 cm, permanent inundation would shift inland up to 25 km (ICEM 2009; Wassmann et al. 2004), and rice yield may be reduced by over 40% (ICEM 2009). Alternately the ADB (2009) estimates that under the scenario of a 1 m sea level rise a 7% decrease in agricultural production would occur, with 10% of the total population directly affected (Office

Table 5.16 Impacts of sea-level rise

Indicator	Occupation	Income	Food	Migration
Cambodia	No direct trans-boundary impacts	No direct trans-boundary impacts	Increase in price of rice from Vietnam and Thailand	Increase in migration from Vietnam into Cambodia
China's Yunnan province	No direct trans-boundary impacts	No direct trans-boundary impacts	No direct trans-boundary impacts	No direct trans-boundary impacts
Southern Lao PDR	No direct trans-boundary impacts	No direct trans-boundary impacts	Increase in prices of rice from Thailand	No direct trans-boundary impacts
Myanmar	No direct trans-boundary impacts	No direct trans-boundary impacts	n/a	No direct trans-boundary impacts
Thailand	Increased wage employment Decreased farming households in Chao Praya River delta	Reduced income from farming in the Chao Pra River delta	n/a	n/a
Vietnam	Increase in wage employment Reduced farming households in the Mekong and Red river deltas	Reduced income from farming in the deltas	n/a	Increase migration to other areas including into Cambodia and Lao PDR

Note: n/a – not available

of the National Target Programme to Respond to Climate Change 2009), and more than one million people in the Mekong delta will be directly affected by sea-level rise by 2050 (Cruz et al. 2007). ICEM (2009) predicts the more modest 20 cm sea level rise is sufficient to increase food scarcity. Total agricultural lands are likely to be reduced, resulting in higher food prices, affecting the whole country because of the delta's importance for national food production (ibid). Table 5.16 summarizes the impacts.

As Vietnam is the second largest rice exporter in the world, the decrease in rice production will also have regional and international consequences. Similarly, the impacts of sea level rise in the Chao Phraya delta would have regional and international impacts. There is no data available quantifying the exact exportation of rice to Cambodia and Lao PDR, but rice from Thailand is sold in all large Lao towns. One of the most important agricultural areas in Thailand is the Chao Phraya River delta (Szuster 2003), implying a large proportion of the Thai rice sold in Lao PDR could be from this part of the country.

Fisheries will also be affected, although reduced catches may not be significant. It is estimated that 2.5 m tonnes per annum of fish are caught in the lower Mekong Basin, of which 30% are caught in Vietnam's Mekong delta (Parry 2007).

The inland intrusion of salt water caused by sea level rise would considerably change the species composition of fisheries, although the change may not be detrimental for overall fish yields (ibid).

According to Warner et al. (2008), flooding is a regular annual occurrence and is an integral part of existing livelihood adaptation strategies in the Mekong delta. However, studies indicate that there is a positive correlation between environmental problems (including flooding) and the magnitude of migration/displacement. Human trafficking into neighbouring areas is one extreme coping strategy used by many families exposed to water related stresses.

Assuming the sea level rise happens now, migration of a large part of the population in the inundated 25 km zone to urban towns such as Ho Chi Minh City would occur, with an attendant change in occupation from farming to wage employment. In 2008, 55% of the 17.2 million people in the Mekong delta worked as farmers (General Statistics Office of Vietnam 2010). The current rates and direction of migration indicate that segments of the affected people would cross borders into Cambodia and Lao PDR. As a result, income from wage labours and remittances will increase.

Realistically however, the 20 cm sea level rise will occur incrementally over the next 20 years. This would mean that job creation in Vietnam would significantly increase, while the growth rate of the working age population would slowdown, requiring labour inputs from outside the inundated areas, counter-balancing emigration to other countries.

3.4 Large-Scale Rubber Plantations

- Rubber plantations affect livelihoods in a number of ways
- Reduced availability of agricultural lands, including grazing lands
- Labour shortages and altered distribution (Baird, undated)
- Increased pressure on common properties, including forest resources and biodiversity
- Altered flow regimes of water resources (Baird undated, Lang 2006) and
- Health impacts caused by the use of chemical herbicides, and pesticides (Baird undated)

Lang (2006) recognized that "the rapid growth of rubber plantations causes large-scale loss of forest resources and watershed destruction, which is particularly important in Lao PDR where rural food security is directly related to forest health".

It is important to distinguish between large and small-scale rubber plantations. Baid (undated) indicates that small holder farming households have the potential to economically benefit from small-scale rubber plantations, while large-scale rubber plantations have a massive and rapid impact on landscapes and livelihoods. According to Table 5.17, assuming large-scale rubber plantations account for 50% of the total estimated expansion, the total increase would be approximately

Table 5.17 Rubber plantation expansion in the wider Mekong Region, ('000 ha)

Year	Unit	Total	Cambodia	Myanmar	Laos	Thailand	Vietnam	Yunnan of China
2000/2002		2,902	50	180	50	2000	412	210
2050		4,500	422	411	194	2244	696	533
Increase		1,598	372	231	144	244	284	323
50% of the increase		799	186	116	72	122	142	162
Average agricultural land per household	ha/hh		0.9	2.4	1.6	3.5	0.63	0.64

Source: Average agricultural land per household

1. Cambodia: Agrifood consulting international. 2005
2. Myanmar: Department of Statistics 2000
3. Lao PDR: Department of statistics, LAO AGRICULTURAL CENSUS, 1998/99
4. Thailand: Pongsrihadulchai 2009
5. Vietnam: Food and agriculture organization of the United Nations and General Statistics Office/Ministry of Planning. 2010
6. Yunnan Province: Cuddy et al. 2003. Note: the term used was average cultivated area per farm

800,000 ha. The largest increase occurs in Cambodia, followed by Yunnan Province. Average agricultural land per household is already limited, particularly in Yunnan Province and Vietnam, about 6.3 ha per household (Table 5.17).

A summary of the impacts of large-scale rubber plantations is provided in Table 5.18. Documentation of large-scale rubber plantations in Lao PDR reveals the loss of land access reduces the livelihoods local farmers (e.g. Thongmanivong et al. 2009). Large rubber concessions have contributed to increased poverty, particularly ethnic minorities in upland areas (Chamberlain 2007). The loss of land-based assets and attendant food and income resources that has pushed local farmers to become wage labourers observed in Lao PDR (Thongmanivong et al. 2009) is likely to also arise in other wider Mekong Region areas. This suggests that an increase of rubber plantations will increase the number of wage labourers and decrease the number of subsistence farmers in all Mekong countries.

Resilient capacities of the affected population who remain subsistence farmers would be weakened due to the loss of important household assets such as livestock and access to natural resources. Large-scale rubber plantations are likely to cover a large proportion of grazing lands, reducing the carrying capacity for livestock. The quality of local and regional groundwater and surface water is likely to decline due to the use of chemicals such as herbicides, unless appropriate mitigation measures are taken. In turn, this will impact on terrestrial and aquatic ecosystems and as a corollary, income and food security.

As locals have limited livelihoods alternatives, losing access to land and forest resources forces them to relocate as the only survival option (Thongmanivong et al. 2009). Migration might include cross-border migration, mainly from Cambodia, Myanmar, and Lao PDR into Thailand. Alternatively, as observed in Lao PDR, large numbers of Chinese workers dominate the north, and Vietnamese employees (legal and illegal) dominate the south. Most, if not all, concessions for large-scale plantations are assigned to foreign investors. Foreign companies are usually accompanied by foreign managers and workers (Baird undated).

Table 5.18 Impacts of large-scale rubber plantations

Indicator	Occupation	Income	Food	Migration
Cambodia	Decrease in rice farming households Increase in wage labourers	Increase in income from rubber plantation Decrease income from rice farming and collection of natural resources	Increase in purchased food items Reduced own produced food	Increase migration into Thailand
China's Yunnan province	Decrease in rice farming households Increase in wage labourers	Increase in income from rubber plantation	n/a	Increased migration into Cambodia, Myanmar and Lao PDR
Lao PDR	Decrease in rice farming households Increase in wage labourers	Increase in income from rubber plantation Decrease income from farming and collection of natural resources	Increase in purchased food items Reduced own produced food	• Increase migration from China in the North and from Vietnam in the south
Myanmar	Decrease in rice farming households Increase in wage labourers	Increase in income from rubber plantation Decrease income from farming and collection of natural resources	Increase in purchased food items Reduced own produced food	Increase migration from China
Thailand	Decrease in rice farming households Increase in wage labourers	Increase in income from rubber plantation Decrease income from farming and collection of natural resources	Increase in purchased food items Reduced own produced food	Increase migration from Cambodia, Myanmar, and Lao PDR
Vietnam	Decrease in rice farming households Increase in wage labourers	Increase in income from rubber plantation Decrease income from farming and collection of natural resources	Increase in purchased food items Reduced own produced food	Increase migration into Cambodia and northern Lao PDR

Note: n/a – not available

The increase in small-scale rubber plantations would imply a reduction in rice farming as a primary occupation. This would also imply a drop in average household income from rice farming, offset by an increase in income from rubber production.

3.5 Large Scale Transport Infrastructure, Specifically the Regional Railway Lines

Although transport infrastructure development improves livelihoods outcomes through better access to mainstream development and social services, it does initiate long term changes in land use resulting from greater accessibility to rural natural resources and encouraging development projects at the expense of natural ecology and indigenous people's livelihoods (e.g. Fouracre 2001). Development of transport infrastructure, inevitably results in large-scale displacement and resettlement of local communities, particularly ethnic minorities in rural areas, who lose their land and other livelihood sources and find it difficult to adapt to the new environment. Table 5.19 summarizes impacts of large-scale infrastructure. The regional railway lines in all countries will further attract and facilitate large scale projects, including resource extraction such as mining, creating thousands of jobs, during construction and operation. However, 75% of construction contracts in Lao PDR have been assigned to Chinese companies and construction settlements comprise up to 20,000 Chinese workers, often in sparsely populated areas previously inhabited by local ethnic communities (Caouette et al. undated).

From a regional perspective, the railways are likely to create more jobs during and after construction. Wage employment and income would increase, reducing the proportion of income from self-employment in farming. Purchased food as opposed to own produced items would increase.

The most far-reaching impact of the railway is that it would facilitate affordable and expedient labour movement, previously considered difficult and prohibitively costly, making domestic and cross-border migration more attractive and feasible. This is particularly favourable for the poor and low-skilled people from Cambodia, Myanmar and Lao PDR migrating into Thailand (and Vietnam by 2030), while skilled labours from China, Thailand, and Vietnam will migrate into Cambodia, Lao PDR, and Myanmar.

3.6 Large-Scale Mining Activities, Specifically Bauxite Mine in the Bolaven Plateau

Mining potential is high in Lao PDR, with 511 mineral deposits and 269 exploitation projects recorded in early 2010, contributing 85% of the total export and 18% of Lao PDR GDP in 2009 (Bounhnaphalom 2010). Mining is currently not conducted with effective environmental management, mitigation, and rehabilitation, with observed impacts of severe loss of vegetation, water pollution, extensive soil erosion and siltation, and toxic metal discharge (ibid).

3.6.1 Livelihoods in Champasak Province

Coffee plantations are the most important permanent crop in Lao PDR, mainly grown on the Bolaven plateau, Champasak province, where bauxite mining is either

Table 5.19 Wider Mekong region impacts of large-scale regional railways

Indicator	Occupation	Income	Food	Migration
Cambodia	Increase wage labourers	Increased income from wage employment	Decrease own produced food items Increase purchased food items	Increased migration into Thailand
China's Yunnan province	Increase wage labourers	Increased income from wage employment and remittances	n/a	Increased migration into Cambodia, Myanmar, and Lao PDR
Lao PDR	Increase in wage labourers	Increased income from wage employment	Increase purchased food items Decrease own produced food items	Increased migration from China into Lao PDR
Myanmar	Increased wage labourers	Increase income from wage employment	Decrease own produced food items Increase purchased food items	Increased migration into Thailand
Thailand	Increased wage labourers	Increase income from wage employment	Decrease own produced food items Increase purchased food items	Increased migration from Cambodia, Myanmar, and Lao PDR
Vietnam	Increased wage labourers	Increase income from wage employment and remittances	Decrease own produced food items Increase purchased food items	Increased migration into Cambodia and Lao PDR, But increased migration from these countries by 2030

Note: n/a – not available

located or proposed. At least 138 km^2 (or 13,800 ha) of the Bolaven plateau have been ceded for mine explorations (Ord River Resources Limited 2006). This is considered bigger than any natural resource projects currently undertaken in Lao PDR, including the Sepon gold and copper mine (ibid). By 2000, about 41,000 ha of coffee had been planted by 24,000 coffee growers (Steering Committee for the Agricultural Census Department of Statistics 2000), or 136,800 persons.[5] In Bachiang district, Champasack province, coffee is the main crop for about 15,000 households, the main source of income for more than 80% of households in the

[5]According to Steering Committee for Census of Population and Housing, 2005, average household size in Champasak is 5.7.

Table 5.20 Some characteristics of Champasack Province

Distribution of occupation by sector in 2005 (%)		
	Total active population	305,407
	Farming	62.90%
	Fishing	0%
	Livestock	0.10%
	Mixed farming	17.30%
	Non-farming	19.60%
	Total	100%
Annual household income in 2007/08 (000 kip)		
	Grain	7,222
	Vegetables and fruits	1,920
	Meat	1,843
	Fish	1,681
	Forestry	170
	Other	1,168
	Total	14,004
Share of income in 2007/08 (%)		
	Grain	51.6%
	Vegetables and fruits	13.7%
	Meat	13.2%
	Fish	12.0%
	Forestry	1.2%
	Other	8.3%
	Total	100%

Source: Department of statistics, Lao PDR

area, with a total plantation area of 45,000 ha, or 30–40% of the total land area (Lazarus 2009). In 2007/2008, average annual income per household was 14 million kip (about US$1,647) in Champasak, representing US$4.5 per household/per day, or US$0.80 per person per day; less than the national poverty line of US$1.00 per person per day. More than 80% of the total income is derived from agriculture, fishery, and forestry Table 5.20.

3.6.2 Impacts

Impacts of large-scale mining activities, specifically bauxite mine in the Bolaven plateau are provided in Table 5.21. Benefits brought about by mining activities to local communities would include development of social infrastructure, agriculture, forestry, small-sized industry, and services, leading to more jobs and positive change (Lazarus 2009). It is likely that employment opportunities created by the mines would benefit many people from the surrounding countries, likely China. Very few local people, if any from ethnic minority groups, are likely to benefit from mining employment opportunities (Lazarus 2009).

Table 5.21 Large-scale mining activities, specifically bauxite mining in the Bolaven Plateau

Indicator	Occupation	Income	Food	Migration
Cambodia	No direct trans-boundary impacts	Negligible	Negligible	No direct trans-boundary impacts
China's Yunnan province	Increased wage employment	Increase income from remittances.	n/a	Increased migration into Lao PDR
Lao PDR	Increase wage employment	Reduced income from agriculture, particularly coffee. Increase income from remittances.	Reduced income from NTFPs (non-timber forest products)	Increased urban–rural migration. Increased migration into Thailand
Myanmar	No direct trans-boundary impacts	No direct trans-boundary impacts	No direct trans-boundary impacts	No direct trans-boundary impacts
Thailand	No direct trans-boundary impacts	No direct trans-boundary impacts	No direct trans-boundary impacts	Increased migration from Lao PDR
Vietnam	No direct trans-boundary impacts	Increased income from remittances	No direct trans-boundary impacts	Increased migration into Lao PDR. Increased migration from Lao PDR, by 2030

Note: n/a – not available

Based on the baseline data provided in the previous Section, assuming the total area of the Bolaven plateau is allocated for bauxite exploitation, 136,800 people would be directly affected. This is almost equal to half of the provincial population. Average income is already less than $1.00 per day. Yet, most of the income (more than 80%) is derived from the agriculture, fishery, and forestry derived from proposed mining areas.

Replacement and adaptation would be the foreseen solutions. Change from farming to wage employment is likely, resulting in the increase in wage income and purchased food. The change would also involve cross-border migration. At present condition, cross-border migration would be into Thailand, while under the 2030 scenario, another choice would be into Vietnam. As the mining will be exploited mainly by foreign companies, notably from China and Vietnam, it is expected that the next two decades will witness an increasing number of Chinese workers in Lao PDR.

Impacts of the mines are expected to be significant and well beyond the district and provincial boundaries. Thousands of people rely on the Sekong River for their livelihoods, threatened because of potential tailing contamination and educed river flow regimes (Lazarus 2009). Concerns over groundwater and surface water quality due to discharge of toxins have already been raised in Cambodia (ibid).

4 Cumulative Assessment

Table 5.22 summarises the forecast consequences on wider Mekong Region livelihood of impending development decisions discussed in the Chapter. Cumulatively, more than half of the population in the LMB countries, mainly rural people, would experience a decline in household food and income because of the impacts of the six impending decisions: the decline in fish, crops/vegetables (related to the loss of river-bank gardens and agricultural lands), OAAs (related to the loss of wetlands and biodiversity) and non-timber forest products (NTFPs) (related to the loss of forests).

The Mekong delta will be affected the most, not just because of its biggest proportion of resource users in the Mekong corridor, but also because of the cumulative interactions of a number of drivers: sea-level rise and other upstream developments, manifest as a 40% decrease in fish yield due to dams and 40% decrease in rice yield due to the 20 cm sea-level rise. Accumulation of herbicides by the large-scale rubber plantation, and industrial pollutants by bauxite mines would pose significant reductions in water quality in the delta, and hence threats to aquatic life and public health.

In Cambodia, the majority of rural households would be more vulnerable because of limited livelihood alternatives, particularly in the Tonle Sap area, where close to 30% of the population under the SIMVA sites do not have a second occupation or livelihood activity; they are singularly dependent on fisheries for livelihoods. The fish catch would decline from more than 4 kg/day to below 1 kg/day as a result of the combined effects of dams, water diversion and bauxite mining. The water diversion and bauxite mines would cause reduced dry-season flow, bringing about adverse impacts on fisheries (e.g. Johnston et al. 2009).

In the Lao PDR, more than 50% of the rural population would be affected. Population of the 15 km corridor accounts for 53% of the national population in 2007 (3.1 million people). Adding rubber plantations (about 260,000 people), and water diversion (about 250,000 people), the affected population is approximately 3.6 million people or 61% of the total population. Additionally some households are likely to be affected by more than one driver. Aggregating the impact of all six regional drivers would negatively affect approximately 65% of the total Lao PDR population. Food availability and income for the majority of these people would decrease compared to current conditions if all the changes were to occur. Average income per capita per day as shown by the case of Siphandone would fall below $1.00 per day unless other sources are made available are sufficient to compensate for the losses. Self-reliance (consumption of own production as opposed to purchased food items) would significantly decline. Malnutrition is likely to increase.

In Thailand, opportunities created by water diversion, rubber plantations and regional railways are likely to increase commercial farming, with an attendant increase in wage employment and reduced subsistence farmers, compared to conditions prevailing in the year 2000. In the Mekong corridor, northeast Region, the decrease in fish availability would affect almost all households. SIMVA data indicate

Table 5.22 Summary of impacts

Decisions	Impacts by projects	Cumulative impacts (interactions of changes)
1. Mainstream dams	• Occupation: No change at present conditions, but by 2030 the proportion of fishing household would decline in all countries • Income: reduced proportion of income from fish in all LMB countries, but might increase income from rice productivity in the Mekong delta (salinity reduced in the dry season by higher flow) • Food: Reduced proportion of food from fish in all countries, but would increase purchased food in the delta (due to increase income from rice) • Migration: Plausible increase migration from Lao PDR and Cambodia into Thailand at present condition if all the 12 are built. Under 2030 scenario, Vietnam would be another destination for the labour migration from Cambodia and Lao PDR	Occupation: • Farming households in NE Thailand would decrease due to opportunities for expansion of commercial farming, as a result of regional water diversion, expansion of rubber plantation, and sea level rise in the Chaopraya river basin • Farming households in the Mekong delta would decrease due to the combined effects of 20 cm sea level rise, and water diversion. Although the mainstream dams can increase dry season flow, reducing salt water intrusion, increased water diversion reduces net downstream flow • Fishing households in the Mekong delta, Cambodia, and Lao PDR would decrease as a result of combined effects of mainstream dams, water diversion, and mining
2. Water diversion	• Occupation: Increased in commercial farming, while decrease in subsistence farming households (or small farmers). Increase wage employment in NE Thailand • Income: increase in income from wage/salaries, noticeably in NE Thailand. Plausible increase in income from remittances in Cambodia, southern Lao PDR, and Mekong delta • Food: reduced own produced food in all countries while purchase food increases in NE Thailand • Migration: likely more labour from Cambodia, Myanmar, and Lao PDR	Income: • Decrease in the proportion of income from fish in all countries (as a results of dams, water diversion, rubber plantation, mining) • Reduced income from farming in the Mekong delta because of sea level rise (loss of agricultural lands and increase salinity intrusion), water diversion, rubber plantation, and mining, reducing the ability to battle the salinity intrusion in the Mekong delta • Increased income from wage employment in all countries, due to mainstream dams, water diversion, sea-level rise, large-scale rubber plantation, regional railway, and bauxite mining • Increase income from remittances due to the increase labour migration as a result of all the six decisions

(continued)

Table 5.22 (continued)

Decisions	Impacts by projects	Cumulative impacts (interactions of changes)
3. Sea-level rise	• Occupation: slight increase in wage employment in Cambodia • Income: likely no significant trans-boundary impacts • Food: likely no significant trans-boundary impacts • Migration: likely increase migration into Cambodia and Lao PDR. Under the 2030 scenario, the migration from Vietnam into Cambodia and Lao PDR will slow down due to the increased job creation in Vietnam and slowdown the growth of labour force in Vietnam	**Food:** • Decreased the proportion of food of the poor from capture fisheries in all LMB countries (as a results of dams, water diversion, mining) • Decreased the proportion of own produced food in all LMB countries (as a result of mainstream dams, water diversion, mining, rubber plantations) • Increase the proportion of purchased food in all countries
4. Large-scale rubber plantations	• Occupation: reduced rice farming households, while increased in rubber plantation households in all countries • Income: increase income from rubber plantations in all countries • Food: increased purchased food items and reduced own produced/gathered food • Migration: increase migration from Lao PDR into Thailand, and from China and Vietnam into Lao PDR	**Migration:** • Increase migration from Cambodia and Lao PDR into Thailand (as a result of mainstream dams, rubber plantations, transportation, water diversion), and Myanmar into Thailand due to the regional railway and large-scale rubber plantation • Increase migration from Thailand, Vietnam, and China into Cambodia and Lao PDR (as result of mainstream dams, mining, railway, and rubber plantations)
5. Large-scale transportation (railway)	• Occupation: wage employment would increase regionally, farming population would reduce • Income: increase in regional income from wage employment • Food: increased purchased food • Migration: Increase migration across countries (during and after construction)	
6. Large-scale mining	• Occupation: increase wage employment in Lao PDR • Income: increase wage employment and remittances Food: likely no direct trans-boundary impacts Migration: increase cross-border migration (during construction and operation)	

that 90% of total households consumed some fish in the week before the survey, and fish contributes 11% of the total households calorie intake. However, fish comprises 0.4% of the total cash income. For this reason, the impacts of a predicted decline in fish would not significantly affect cash income in this part of Thailand. Migration from neighbouring countries, particularly from Cambodia, Lao PDR, and Myanmar into Thailand would increase.

In Myanmar and Yunnan Province, most of the impacts would be catalysed by a regional railway and large-scale rubber plantations. The loss of forest resources, agricultural lands and more affordable transportation brought about by the railway would be an important push factor for increased migration into Thailand. In turn, this will result in an increase in the income from remittances. In Yunnan Province, the majority of impending developments would create an environment promoting the migration of more skilled labour to Cambodia, Myanmar and Lao PDR. These include construction and operation of dams, mining, railway, and large-scale rubber plantations. Income from remittances to Yunnan Province would increase significantly.

Combined with the continued differences in labour force availability, differential rates of economic growth of wider Mekong Region countries, and the decline in natural resources, the regional railway would strengthen the increasing pattern of migration from Cambodia, Myanmar and Lao PDR into Thailand. Factors contributing to migration in the wider Mekong Region are attributed to:

- The relationship between economic growth (Mekong Migration Network and Asian Migrant Centre 2009)
- Population growth[6]
- Lack of opportunities (Lewis et al. 2010)[7]
- The level of payment, poor quality of social services (Caouette et al. undated)[8]

[6]The Thai *labour supply is growing slower than the creation of new jobs*, resulting in a demand for inward migrants (Lewis et al. 2010; Mekong Migration Network and Asian Migrant Centre 2009). Thailand is increasingly experiencing the challenges created by an *aging population, population growth rates in the poorer countries in the sub-region are high*, between two and three times higher than that of Thailand, and an increasing share of people of working age (Revenga et al. 2006). Thailand becomes a labour receiving country, because it passed the peak in its population growth rate some time ago, while Vietnam has just passed the peak recently, hence still has excess labour supply relative to its economy because most of it population is of working age (Lewis et al. 2010).

[7]It is not always true that frequent migration of migrants in the wider Mekong Region is due to the opportunity to earn higher wages, but because of the *lack of opportunity* in their home countries. Wages in Thailand are higher than in other countries of the wider Mekong Region, but are not many times higher (Lewis et al. 2010). Nevertheless, *higher incomes and wage levels together with the demand for low-skilled labour* in Thailand has attracted many poorly trained rural people, from Cambodia, Lao PDR, and Myanmar (Caouette et al. undated).

[8]*Poor quality and limited services*, especially in health and education in poorer countries are major factors behind the creation of an unskilled and poor workforce, a factor for migration to Thailand, where services, although difficult to access for migrants, are at least of higher quality and available (Caouette et al. undated).

- Differences between countries in demographic structure (Revenga et al. 2006)
- Decline in natural resources (ICEM 2010; Hall and Bouapao 2010)[9] and
- Impacts of flooding (Warner et al. 2009)

By 2030, Vietnam may also become a receiving country for migration and Myanmar may still be a source country for some time to come (Lewis et al. 2010). With opportunities created by large scale investment and infrastructure projects, there will be an increase in migration from China into Cambodia, Myanmar and Lao PDR, and from Thailand and Vietnam into Cambodia and Lao PDR.

5 Conclusion

With a focus on rural communities, this Chapter has evaluated the likely transboundary consequences of the six drivers affecting household livelihoods, Relying primarily on secondary data, the analysis indicates that if the proposed developments eventuate, adverse effects are likely to outweigh benefits. Despite beneficial intentions, the future income and food conditions of millions of people in the wider Mekong Region will deteriorate compared to their current livelihood status, unless sufficient compensatory measures are undertaken. Currently, a high proportion of the wider Mekong Region rural population are either living under the poverty line of US$1.00 per day or are just above it. Implementation of all development projects will increase the proportion of the population living in impoverished conditions. There is a trend of shifting from subsistence farming to wage employment.

Overall, the proportion of household income from natural resources such as fish and NTFPs is likely to decrease, while the average income from wage employment and remittances is likely to increase. This also suggests a diminished reliance on subsistence food production particularly in Lao PDR, where the proportion of subsistence agriculture accounts for about 65% of food consumed in rural areas (with road access) and more than 75% in rural areas without road access (Department of Statistics 2009). This implies that access to natural resources for nutritional purposes will decline. Rural farmers and fishers will be most affected, but non-farming households and urban people will experience indirect effects through increasing food prices as a result of the decline in natural resources.

Migration from Cambodia, Myanmar, and Lao PDR into Thailand will increase as a corollary of the predicted decline in natural assets in concert with the regional railway and other potential development decisions. Vietnam might see migration from Cambodia and Lao PDR by 2030. The next two decades are also likely to witness an increase in skilled labour from China, Thailand, and Vietnam into Cambodia, Myanmar, and Lao PDR.

[9] Rural–urban migration is driven by a declining natural resources base in upland areas and increasing employment opportunities in lowland and urban areas (ICEM 2010).

Generally data are not comparable across wider Mekong Region countries because of differences in timing, definitions, scales and methodologies (MRC 2003). Specific data enabling interrogation to satisfy the need for livelihoods assessment at the regional level is limited. The capacity to decompose available livelihood data to reveal for example income and food sources and the reliable assessment of responses to specific causal factors are not available for some countries. Some wider Mekong Region country data are available at higher resolution such as case studies. Others are at a coarser scale such as regional or national levels. The lack of reliable and commensurate data to appraise the relationship between livelihoods and natural systems introduces an additional knowledge gap and constraint to evidence based policy to improve wider Mekong Region livelihoods.

References

ADB. 2007. *Developing communications for ethnic groups*. Manila, Philippines.

ADB. 2009a. *Water vital for Vietnam future*. Manila, Philippines: Asian Development Bank.

ADB. 2009b. Asian Development Bank & Myanmar: Fact sheet. Manila, Philippines.

ADB. 2010. Asian Development Bank and fact sheet.

Adger, W.N. 1999. Social Vulnerability to climate change and extremes in Coastal Vietnam. *World Development* 27(2): 249–269.

Agrifood Consulting International. 2005. Project brief series: Cambodia agrarian structure study. www.agrifoodconsulting.com/ACI/uploaded.../brief_35_45806374.pdf. Accessed 26 Feb 2011.

Ahmed, M., H. Navy, L. Vuthy, and M. Tiongco. 1998. *Socioeconomic assessment of freshwater capture fisheries in Cambodia: Report on a household survey*, 186. Phnom Penh: Mekong River Commission.

Amornsakchai, S., P. Annez, S. Vongvisessomjai, S. Choowaew, P. Kunurat, J. Nippanon, R. Schouten, P. Sripapatrprasite, C. Vaddhanaphuti, C. Vidthayanon, W. Wirojanagud, and E. Watana. 2000. *Pak Mun Dam, Mekong River Basin, Thailand. A WCD case study prepared as an input to the World Commission on Dams, Cape Town*. www.dams.org.

Baird, I.G. undated. Land, rubber and people: Rapid agrarian changes and responses in Southern Lao PDR.

Baird, I., Z. Hogan, B. Phylaivang, and P. Moyle. 2001. A communal fishery for the migratory catfish pangasius macronema in the Mekong River. *Asian Fisheries Science* 14: 25–41. Asian Fisheries Society, Manila, Philippines.

Balzer T., P. Balzer, and S. Pon. 2005. Traditional use and availability of aquatic biodiversity in rice based ecosystems. Kampong Thom Province, Kingdom of Cambodia. M. Halwart, D. Bartley, and H. Guttman. (ed.) In *Aquatic biodiversity in rice-based ecosystems. Studies and reports from Cambodia, China, Lao People's Democratic Republic and Viet Nam.*

Basin Development Plan. 2006. Basin development plan programme Phase 2 2006–2010. Programme Document, Final Version. MRC Vientiane Lao PDR.

Basin Development Plan Programme, Phase 2. 2010. Assessment of Basin-wide Development Scenarios. Draft Main Report. MRC Vientiane LAO PDR

Bouapao, L., and D. Hall. 2010. Water resources related livelihoods and food security. In *State of the basin report 2010*, ed. R. Taylor and H. Bach, 47–53. Vientiane, Lao PDR: Mekong River Commission. ISBN 978-993-2080-57-1.

Bounhnaphalom, E. 2010. Mineral development in Lao PDR. Presentation for the international conference on mining "staking a claim for Cambodia", Phnom Penh, May 26–27 2010.

Caouette, T., R. Sciortion, P. Guest, and A. Feinstein. Undated. Labour Migration in the Greater Mekong Suregion.

Chamberlain, J.R. 2007. *Participatory poverty assessment (2006). Lao People's democratic republic*. Vientiane: National Statistics Center and Asian Development Bank. 100 pp.

Cruz, R.V., H. Harasawa, M. Lal, S. Wu, Y. Anokhin, B. Punsalmaa, Y. Honda, M. Jafari, C. Li, and N. Huu Ninh. 2007. Asia. Climate change 2007: Impacts, adaptation and 0Vulnerability. In *Contribution of working group II to the fourth assessment report of the intergovernmental panel on climate change*, ed. Parry M.L., Canziani O.F., Palutikof J.P., van der Linden P.J., and Hanson C.E., 469–506. Cambridge: Cambridge University Press.

Cuddy, M.P., H. Liu, and S. Steele. 2003. A socio-economic study of Keland Village in Yunnan province in relation to uptake by farmers of improved management practices on red soils in China. Working paper no. 71. Galway: Department of Economics, National University of Ireland.

Davies, S. 1996. *Adaptable livelihoods: Coping with food insecurity in the Malian Sahel*. London: Macmillan.

Department of Statistics. 2000. *Lao agricultural census, 1998/99: Highlights*. Vientiane: Steering Committee for the Agricultural Census.

Department of Statistics. 2006. *Lao PDR population and housing census 2005*. Vientiane: Department of Statistics, Ministry of Planning and Investment.

Department of Statistics. 2009. The household of Lao PDR: Social and economic indicators. survey results on expenditure and consumption of household 2007/2008. LECS 4.

Department of Statistics, Ministry of Housing and UNICEF. 2006. Lao PDR maternal and child nutrition survey, Department of Statistics of Ministry of Planning and Investment, Hygiene and Prevention Department of Ministry of Health and UNICEF, Vientiane.

DFID. 1999. Sustainable livelihoods guidance sheets. Department for International Development. http://training.itcilo.it/decentwork/staffconf2002/presentations/SLA%20Guidance%20notes%20Section%202.pdf. Accessed 21 Oct 2010.

Ellis, F. 1998. Household strategies and rural livelihood diversification. *Journal of Development Studies* 35(1): 1–38. Accessed 31 Dec 2010.

FAO. 2005. Increasing the contribution of small-scale fisheries to poverty alleviation and food security, *FAO technical guidelines for responsible fisheries*. No. 10. Rome: FAO.

FAO. 2010. National gender profile of agricultural households, 2010: Report based on the 2006 rural, agriculture, and fishery census. www.fao.org/docrep/012/k8499e/k8499e00.pdf. Accessed 26 Feb 2011.

Fouracre, P. 2001. Transportation and sustainable rural livelihoods. Rural Travel and Transport Program 2001. TRL Limited Crowthorne. UK.

General Statistics Office of Vietnam. Monthly Statistical Information October 2010 http://www.gso.gov.vn/default_en.aspx?tabid=467&idmid=3&ItemID=9874. Accessed 18 Nov 2010.

GOL and UNDP. 2008. *Millennium development goals: Lao PDR progress report*. Vientiane: Government of Lao PDR and UNDP.

GOL and UNDP, 2009. *4th Lao PDR National Human Development Report*. Vientiane: Government of Lao PDR and UNDP.

GSO. 2010. Results of the Survey on household living standard 2008 http://www.gso.gov.vn/default_en.aspx?tabid=515&idmid=5&ItemID=9647. Accessed 25 Nov 2010.

Hall, D., and L. Bouapao. 2010. *Social impact monitoring and vulnerability assessment in the Mekong corridor: Report on a regional pilot study. MRC Technical Paper No. 30*. Vientiane: Mekong River Commission, 146 pp.

Halwart, M. 2008. Biodiversity, nutrition and livelihoods in aquatic rice-based ecosystems. *Biodiversity Journal of Life on Earth* 9: 36–40.

Hortle K.G., and U. Suntornratana. 2008. Socio-economics of the fisheries of the lower Songkhram River Basin, northeast Thailand. MRC technical paper no. 17. Vientiane: Mekong River Commission, 85 pp-.

ICEM. 2009. Forum report: Mekong delta climate change forum 2009. For office of Government, MONRE, Australian Embassy, Danish Embassy, World Bank and Asian Development Bank, 12–13 Nov 2009.

ICEM. 2010. *MRC strategic environmental assessment (SEA) of hydropower on the Mekong mainstream*. Hanoi, Viet Nam. Mekong River Commission http://www.mrcmekong.org/ish/SEA.htm. Accessed 21 Oct 2010.

Idea International Institute, Canada and IHLCA Project Technical Unit, Myanmar. 2007. Integrated household living conditions survey in Myanmar: Poverty profile www.mm.undp.org/UNDP_Publication…/IHLCA-Poverty%20survey.pdf. Accessed 26 Feb 2011.

Johnston, R.M., C.T. Hoanh, G. Lacombe, A. Noble, V. Smakhtin, D. Suhardiman, S.P. Kam, and P.S. Choo. 2009. Scoping study on natural resources and climate change in Southeast Asia with a focus on agriculture. Report prepared for the Swedish International Development Cooperation Agency by the International Water Management Institute. 118. doi:10.3910/2010.201.

Lang, C. 2006. Lao PDR: Vietnamese companies set up rubber plantations in the south. MRC Bulletin, Issue 113 – December 2006. http://www.wrm.org.uy/bulletin/113/viewpoint.html#Lao PDR. Accessed 06 Dec 2010.

Larson, S. 2010. *Collection of primary data for the Mekong futures project*: Document version 4 of July 2010.

Lazarus, K.M. 2009. In search of aluminium: China's role in the Mekong Region. Heinrich Boll Stiftung, WWF and International Institute for Sustainable Development.

Le, T.V.H., H.N. Nguyen, E. Wolanski, T.C. Tran, and H. Shigeko. 2007. The combined impact on the flooding in Vietnam's Mekong delta of local man-made structures, sea-level rise, and dams upstream in the river catchment. *Estuarine, Coastal and Shelf Science* 71: 110–116.

Lewis, D.R., N. Lekfuangfu, I.V. Solorano, F. Soda, and C. Natali. 2010. *Forecasting gration flows: The relationships among economic development, demographic change and migration in the Greater Mekong Subregion*. Bangkok: ADB.

Network, Mekong Migration, and Asian Migrant Centre. 2009. *Migration in the Greater Mekong Subregion. Annotated bibliography*, 4th ed. Hong Kong: Mekong Migration Network and Asian Migrant Centre.

Messerly, P., A. Heinimann, M. Epprecht, P. Souksavath, T. Chanthalanouvong, N. Minot, editors. 2008. *Socio-economic Atlas of the Lao PDR- an analysis based on the 2005 population and housing census,* Swiss National Center of Competence in Research (NCCR) North–south, University of Bern, Bern and Vientiane: Geographica Bernensia.

Meusch E., J. Yhoung-Aree, R. Friend, and S. J. Funge-Smith. 2003. The role and nutritional value of aquatic resources in the livelihoods of rural people – a participatory assessment in Attapeu Province, Lao PDR. FAO Regional Office Asia and the Pacific, Bangkok, Thailand, Publication No. 2003/11.

MOFA. 2007. *Agriculture statistics yearbook 2006*. Vientiane: Ministry of Agriculture and Forestry.

MOFA. 2008. *Agriculture statistics yearbook 2007*. Vientiane: Ministry of Agriculture and Forestry.

Mollot, R., C. Phothitay, and S. Kosy. 2003. Hydrology, habitat and livelihoods on the floodplains of southern Lao PDR. *Proceedings of the 6th technical symposium on Mekong Fisheries*, Pakse Lao PDR 26–28 Nov 2003.

MRC. 2003. *State of the basin report 2003*. Mekong River Commission, Phnom Penh, 300 pages. ISSN: 1728:3248

MRC. 2010. *State of the basin report 2010*. Vientiane, Lao PDR: Mekong River Commission.

National Statistical. 2004. *Lao expenditure and consumption Survey 2002/03 (LECS 3), National Statistics Centre*. Vientiane: Committee of Planning and Investment.

National Statistical Office. 2007. Table compiled by National Statistical Office: The labour force survey 2007. http://service.nso.go.th/nso/nso_center/project/search_center/23project-en.htm. Accessed 24 Dec 2010.

National Statistical Office. 2008. *Statistical year book 2008*. Bangkok: National Statistical Office.

National Statistical Office. 2010. Thailand. http://web.nso.go.th/eng/en/stat/socio/soctab3.htm. Accessed 22 Dec 2010.

Office of the National Target Programme to Respond to Climate Change. 2009. National target programme of Vietnam to respond to climate change. PowerPoint Presentation. Mekond Delta Climate Change Forum. 12–13 Nov 2009.

Ord River Resources Limited. 2006. Lao PDR Bolaven Plateau Bauxite Project Update.

Osborne, M. 2010 (October 26). Dam busting on the Mekong. Business spectator. http://www.businessspectator.com.au/bs.nsf/Article/Mekongs-hydropower-spillover-pd20101026-AKTJL?OpenDocument. Accessed 27 Oct 2010.

Parry, M.L. 2007. Climate change 2007: Impacts, adaptation and vulnerability: contribution to the IPCC WGII fourth assessment report. Intergovernmental panel on climate change.

Pech, S., and K. Sunada. 2008. Population growth and natural resources pressures in the Mekong River Basin. Ambio, 373, no. 3 (May 2008). Royal Swedish Academy of Science 2008.

Pham Trong Thinh. 2010. *The findings and recommendations for social impact monitoring and vulnerability assessment (case study in fresh water corridor of Mekong River in the Mekong delta Vietnam): Final report*. Vietnam National Mekong Committee.

Podger, G., R. Beecham, D. Blackmore, C. Perry, and R. Stein. 2004. Modelled observations on development scenarios in the lower Mekong Basin: Mekong regional water resources assistance strategy. World Bank.

Pongsrihadulchai, A. 2009. Use of farmers' registration for agricultural policy implementation in Thailand. Expert consultation on statistics in support of policies to empower small farmers, Bangkok, 8–11 Sept 2009.

Revenga, A., P.Y. Fallavier, J. Larrison, and C.d.P. Nieves. 2006. Labour migration in the Greater Mekong Sub-region Synthesis Report: Phase I.

Sone, P. 2009. Final national technical report – Lao PDR: Social impact monitoring and vulnerability analysis.

Steering Committee for Census of Population and Housing. 2005. National Statistics Centre, Lao PDR.

Szuster, B.W. 2003. *Shrimp farming in Thailand's Chao Phraya River Delta: Boom, bust and echo*. Colombo: International Water Management Institute (IWMI). Sri Lanka.

The Nations. 2010. [Editorial] Think twice before doing irreversible damage. The Nations http://www.koreaherald.com/national/Detail.jsp?newsMLId=20101001000840. Accessed 04 Oct 2010.

The World Bank. 2010. Rural population 2010. The World Bank Group http://data.worldbank.org/indicator/SP.RUR.TOTL.ZS?page=1. Accessed 31 Dec 2010.

Thongbai, P. 2009. Agricultural expansion and intensification. In Louis Lebel, Anond Sindvongs. In *Critical states: Environmental challenges to development in monsoon Southeast Asia*, ed. Lebel Louis, Sindvongs Anond, Chen Chen-Tung Arthur, and Daniel Rajesh, 156–174. Malaysia: Strategic Information and Research Development Centre.

Thongmanivong, S., Y. Fugita, P. Khamla, and V. Thoumthone. 2009. Agrarian land use Transformation in Northern Lao PDR: From Swidden to rubber. Southeast Asian Studies, 47, no. 3, December 2009

UNEP and ADB. 2010. Yunnan Sustainable Development Strategy Study. Yunnan Province, P.R. China http://www.rrcap.unep.org/nsds/brief/Yunnan%20brief.pdf. Accessed 30 Dec 2010.

UNESCAP. undated. Rural poverty in South-East Asia: trends, issues and challenges http://www.unescap.org/pdd/publications/RuralPoverty/ChIII.pdf. Accessed 06 Sept 2010.

Warner, K., C. Ehrhart, and A. de Sherbinin. 2009. *In search of shelter: Mapping the effects of climate change on human migration and displacement*. Geneva: Care International. Switzerland.

Warner, K., O. Dun, and M. Stal. 2008. Climate change and displacement. In Forced migration review, eds. M. Couldrey, and M. Herson, eds. 31 p. 13–14. Oxford Department of International Development, University of Oxford.

Wassmann, R., N.X. Hien, C.T. Hoanh, and T.P. Tuong. 2004. *Climate change*, Sept 2004. Vol. 66 Issue 1/2, p89–107, 19p.

WFP. 2007. Lao PDR: Comprehensive Food Security and Vulnerability Analysis (CFSVA). United Nations World Food Programme and European Union.

WFP. Population. 2010 WFP http://foodsecurityatlas.org/khm/country/sitemap. Accessed 31 Dec 2010.

WHO. 2009. Country context: Demographic. http://www.wpro.who.int/countries/2009/cam/. Accessed 25 Oct 2010.

Chapter 6
Land-Use Change in the Mekong Region

Lu Xing

1 Introduction

This Chapter provides an assessment of the impacts of the six development related changes outlined in Chap. 1 on land-use change across the wider Mekong Region. The wider Mekong Region remains an important food production area, with rain-fed and irrigated farmland, aquatic agriculture and pasture as crucial production systems. On the other hand, forests and wetlands provide critical ecosystem services to agricultural production and society. The wider Mekong region experiences fast economic growth with an average GDP growth rates of around 8% over the past decade. The major driving force is investment from within the wider Mekong region and from other parts of the world. The fast development of hydropower, mining, infrastructure and urbanization are main drivers for land use change. Also the area of commercial tree plantations such as rubber has dramatically increased in the region over past decade. These developments establish huge potential threats to traditional land uses and, thereby, for livelihoods and ecosystem services. Given to the aforementioned concerns, the indicators listed in Table 6.1 have been employed to assess land-use changes.

Forest change refers to aggregate changes in either the area of primary and secondary forests or the conversion of forest into alternate land uses and vice visa. Plantation change indicates increases or decreases in the aggregate plantation area of rubber, coffee and other cash trees. Farmland is defined as the aggregate area of land assigned to the production of annual crops that are reliant on rain fed agriculture. Irrigated farmland refers to farmland where irrigation infrastructure has been constructed to allow for dry-season cropping. The irrigation of farmland is an important indicator of the extent of intensive farming. Irrigated plantation is not included in irrigated farmland. Due to the lack of reliable data, this analysis omits cropping rate as an otherwise relevant indicator. Aquatic resources refer to areas under aquaculture.

A. Smajgl and J. Ward, *The Water-Food-Energy Nexus in the Mekong Region: Assessing Development Strategies Considering Cross-Sectoral and Transboundary Impacts*,
DOI 10.1007/978-1-4614-6120-3_6,

Table 6.1 Indicators to assess land use change

Indicator	Definition
Forestland	Area of primary and secondary forests
Plantation	Area of plantation trees crops such as rubber, coffee and other perennial species
Farmland	Area of rain-fed annual crops
Irrigation of farmland	Area of farmland to be fully irrigated
River bank garden	Area of river garden
Aquatic resources	Area of aquaculture

Table 6.2 The changes in Mekong region land use from 2000 to 2008

Land use in 1,000 km^2	2000	2005	2008
Forested land	1,131.45	1,138.05	1,127.14
Pasture	107.44	97.79	97.76
Farmland	405.39	424.11	433.25

Source: FAO, Yunnan Statistics Year Book 2009

2 Land Use Change in Mekong Region

The total land area of the Mekong Region is 2,334,000 km^2. Forested land is still a dominant land use type, followed by farmland and pasture fallow. At the end of 2008, 1,127,140 km^2 or 48.29% of the total land was classified as forest, 433,250 km^2 or 18.56% as farmland and 97,760 km^2 or 4.15% as pasture (Table 6.2). Water surface, less than 4% of the total land area, is a valuable source of food and protein in the wider Mekong region, especially for people in lowland river basins.

Substantial and rapid land use change has occurred in the region. By 2008 farmland increased by 6.9% if compared to 2000 while forested land decreased by 0.38% and pasture decreased by 9.01% (Fig. 6.1). Cash tree plantations boom since 2005. According to Ziegler et al. (2009), plantations of rubber, coffee and other perennial commercial tree crops have expanded rapidly, estimating that more than 500,000 ha of forestland might have been converted to rubber.

Figure 6.2 illustrates the characteristic pattern of land use change during 2000 and 2008, highlighting differences in land use change between Mekong Region countries. Farmland increased by 61.3% in Yunnan and 42.5% in Lao PDR. Pasture increased 61.3% in Cambodia and decreased 21.3% in Yunnan. Forested land increased 15.2% in Vietnam and 11% in Yunnan and decreased 10.4% in Cambodia, 7.1% in Myanmar, 3.8% in Lao PDR and 0.3% in Thailand. However, forest coverage may be over-estimated in Vietnam and in the Chinese Province of Yunnan. These two countries include commercial tree plantation as forested land. In most countries plantations are not classified as a forest type.

The relevance of irrigation varies among Mekong Region countries (Table 6.3). At the end of 2008, Vietnam had the highest percentage of farmland under irrigation, followed by Thailand and Yunnan. Irrigation was less relevant in Lao PDR, Myanmar and Cambodia, potentially indicating a substantial potential to increase

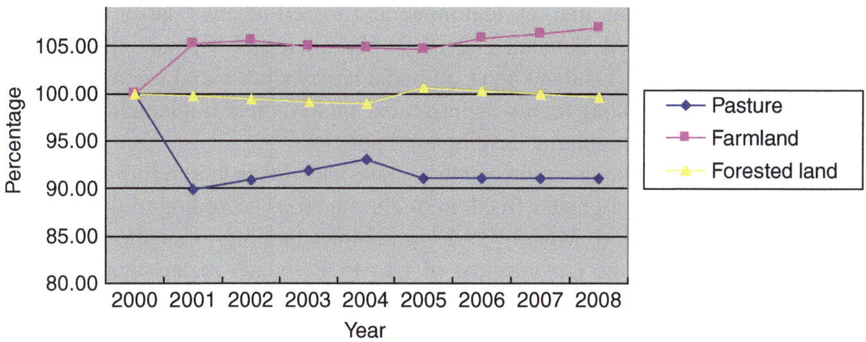

Fig. 6.1 Land use change in the Mekong region (Source: FAO, Yunnan Statistical Year Book 2009)

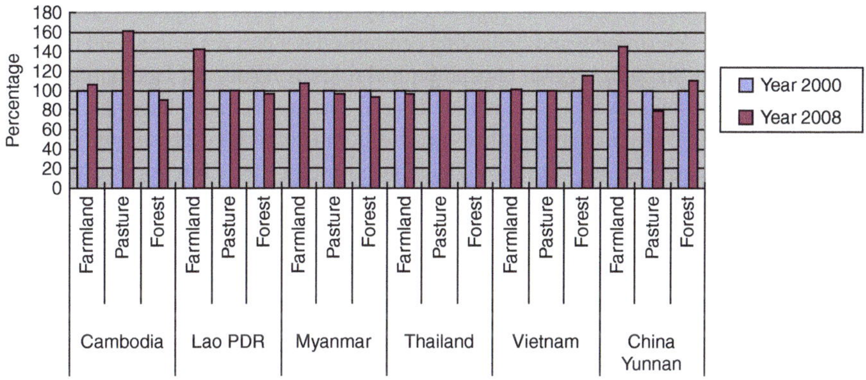

Fig. 6.2 Specific land use change of countries/province 2000–2008 (Source: FAO, Yunnan Statistical Year Book 2009)

Table 6.3 The percentage of total farmland with existing irrigation in the wider Mekong region

Cambodian (%)	Lao PDR (%)	Myanmar (%)	Thailand (%)	Vietnam (%)	Yunnan, PRC (%)
7.0	22.3	19.2	34.1	48.9	30.0

Source: FAO and Yunnan Statistical Year Book 2009

agricultural production through irrigation. Evidently, the Governments of Lao PDR and Cambodia might be able to increase yields by providing irrigation (IWMI 2006). The Cambodia government plans to implement rice double-cropping by expanding irrigation schemes (CDRI 2011).

Land use change results from a series of decisions made by land holders subject to various constraints and partly guided by maximizing return from investing in land as a production factor. Rowcroft (2008) concludes that the causes of land use change are multifaceted: The prices of agricultural and forest products are the most important drivers for land use change. Available technology and infrastructure development (i.e. roads) are also important drivers of land use chance. Changes in land titles are another important driver. Hall et al. (2011) identify four processes that change land title including legal, market, legitimate and forced means (Hall et al. 2011). The Chinese Province of Yunnan, Lao PRD, Myanmar and Vietnam still heavily regulate land use while Thailand and Cambodia have reduced land use regulations. However, all of the Mekong region countries except Myanmar implemented at least some market-oriented economic development policies.

Governments of Mekong region countries have undertaken land reforms since the 1980s. Farmers obtained greater freedom to choose crops according to their conditions and market demand. Model "3+2", a contract farming arrangement, is sincerely encouraged by the Government of Lao PDR, which means that investors introduce and improve funding, market access and new technologies while farmers contribute land and labor in order to meet increasing demand for agriculture products at the regional and international levels. The Government of Thailand has undertaken land reforms by granting more secure land use rights to farmers over the past three decades. This arrangement has entitled farmers to use land as an asset and to better organize production to meet local, regional and global market demands (Lu 2007).

In addition to determining and deploying centralised policies regarding land use regulation, governments in the region play additional roles as holders of land entitlements (and attendant rights) as well as primary land users. The governments have also granted land concessions to diverse private investors for commercial logging and agriculture, industry development, infrastructure and other purposes. Increased foreign investment in Lao PDR and Cambodia has resulted in the substantial conversion of forestland into agriculture and commercial tree plantations. Ziegler et al. (2009) argue that plantations are increasingly replacing forestland and fallowed land of shifting cultivation.

Various protection policies play an important role in land use change in the wider Mekong region. Yunnan has established 158 nature reserves by the end of 2010, with an area of 2.96 million hectares, which equals 7.6% of the total land area (Government of China 2011). The Lao Government plans to increase its forest coverage from less than 41% now to 65% by 2015 in the country's seventh five-year economic plan for 2011–15 (Forest carbon Asia, no date). Governments in the wider Mekong region have also put limits on shifting cultivation for conservation (Fox et al. 2009).

3 Single Factor Impacts

The following synthesis of existing case studies aims to highlight the impacts of the six development related changed outlined in Chap. 1 on land use change in the wider Mekong region.

3.1 Mainstream Dams

The MRC SEA report identifies expanding irrigation, loss of forest, farmland and aquatic resources as the main land use consequences of mainstream dams (MRC 2011). The loss of land is a direct impact of hydropower projects (Table 6.4). The MRC (2001) identifies that the construction of reservoirs, access roads and

Table 6.4 The conversion of land to aquatic resource of dams in mainstream (Units ha)

Land use	Forested land	Rain-fed Farmland	Irrigated Farmland	Riverbank garden	Wetland
Land lost	30,8866	14,957	829	11,966	73,500

Source: MRC (2011)

transmission lines for mainstream dams will result in the loss of approximately 30,886 ha of forest land (mostly degraded forest) and 15,786 ha of agricultural land, inclusive of 829 ha of irrigated land. The report estimates that 11,966 ha of riverbank gardens will be submerged by the reservoirs and wetland loss may amount to about 7% of current wetlands, amounting to 73,500 ha within a 50 km corridor of the Mekong mainstream. The total loss of land to mainstream dams accounts for 410,118 ha, equivalent to 32.8% of farmland in Lao PDR or 1.7% of total land.

The Pak Chom project in Lao PDR, the Ban Koum project in Lao PDR and Thailand and the Latsua project in Lao PDR represent planned multiple purpose projects that are associated with irrigation schemes. The total irrigation area is anticipated to increase by 17,876 ha, primarily for either rice production or cash crops. Irrigation projects associated with hydropower development are likely to improve land productivity and rice production in planned irrigation areas. This may lead to an increase in 77,701 tonnes of rice/year or a value of approximately USD 15.54 million/year (MRC 2011).

The discussion above represents only partial impacts of hydropower development on land use change. The impacts may increase dramatically if the estimated figures included Myanmar and all tributary dams in the region. The potential of hydropower generation in Myanmar is estimated at 39,624 MW (FAO 2004) compared to 30,000 MW (MRC 2010) in the lower Mekong Basin. The potential for land use change due to the development of reservoirs would double if Myanmar develops the full potential of hydropower generation. A recent study concludes, "the completion of 78 dams on tributaries, which have not previously been subject to strategic analysis, would have catastrophic impacts on fish productivity and biodiversity. The results argue for reassessment of several dams planned, and call for a new regional agreement on tributary development of the Mekong River Basin" (Ziv et al. 2012).

3.2 Water Diversion

The 4.84 million hectares of agricultural land in the Isaan region of northeast Thailand account for 45% of Thailand's farmland (Molle and Floch 2008). Currently, irrigated crops are primarily comprised of rice, sugarcane, cassava, tomato and tobacco. This study assumes that 60% of the diverted water from the Nam Ngum Basin will be allocated to irrigate this area. If 50% of water is diverted to rice production, 20% to sugarcane, 20% to cassava and 10% to tomato (Table 6.5), the irrigation area will increase to 279,220 ha by 2015 and to 450,070 ha by 2030, accounting for 5.82% and 9.38% of farmland in Thailand and Isaan respectively. Access to additional water from the proposed diversions will intensify production on current farmland rather than substantially alter land use. Intensified production will provide sufficient impetus to develop additional roads connecting farmland and

Table 6.5 Water diversion and subsequent increase of irrigation area in Issan

Crop	Crop water need (mm/total growing period)[a]	Water use (CM/year/ha)	2015 (9167MCM/ year) '000 ha		2030 (14776MCM/ year) '000 ha	
Rice	400–700×2	11,000	50%	100.00	50%	161.19
Sugarcane	1,500–2,500	20,000	20%	22.00	20%	35.46
Cassava	292–438	3,650	20%	120.55	20%	194.31
Tomato	400–800	6,000	10%	36.67	10%	59.10
Total			100%	279.22	100%	450.07

[a]FAO (1986) and Oguntunde (2005)

agricultural processing plants. Existing farmland will need to be converted to transport infrastructure, although the area will not be significant.

3.3 Sea-Level Rise

Predicted sea-level rise will have a significant impact on land use. The MRC SEA report estimates that 67.50% of the land in the Mekong delta is currently used for agricultural production, 14.64% is built infrastructure and 9.66% wetlands (MRC 2011). If sea level rise leads to a loss of 15% of land in the Mekong delta, there will be 476,850 ha of submerged land including 321,870 ha of farmland and 46,035 ha wet land (Table 6.6). The impact will be doubled or possibly tripled if the assessment includes the deltas of the Red, Chao Phraya and Irrawaddy rivers.

Resettlement of displaced communities will have second order impacts on land use change requiring between 1 and 1.5 million ha of land according to MRC (2011). It seems most likely that existing forestland and pasture land will be converted to farmland. It is likely that communities will be resettled either to urban or to northern rural areas.

3.4 Large-Scale Rubber Plantations

This study estimates that in 2000 a total of about 2.9 million hectares of rubber was planted in the Mekong Region accarding to various online sources, papers and `statistics.[1] Ziegler et al. (2009) estimate more than 500,000 ha of farmland and forest have been converted to rubber plantations between 2002 and 2009. By simply adding the above two figures, the total area of rubber plantation might amount to 3.4 million hectares by 2009. By 2050, the area of land dedicated to rubber and other diversified farming systems may double or triple. According to Fox et al. (2009), approximately 1.6 million ha or 4% of the vegetated area of 2000 would be replaced by rubber

[1]Online sources include www.cria.org.cn, www.rubber.com.cn, www.yunken.com, and Yunnan Statistical Year Book 2009.

Table 6.6 Estimated land use change in the Mekong delta as a result of sea level rise

Land use	%	Area (ha)	Area lost (ha)
Farmland	67.50	2,145,800	321,870
Forestland	0.26	8,200	1,230
Pasture	7.95	252,700	37,905
Wetland	9.66	306,900	46,035
Built-up	14.64	465,400	69,810
Total	100.00	3,179,000	476,850

Source: Calculation based on MRC (2011)

Table 6.7 An estimated rubber expansion among wider Mekong region countries (in 1,000 ha)

Year	Cambodia	Myanmar	Lao PDR	Thailand	Vietnam	China Yunnan	Total
2000	50	180	50	2,000	412	210	2,902
2050	422	411	194	2,244	696	533	4,500

Source: Adapted from Fox et al. (2009)

plantations by 2050. Table 6.7 represents a simulation of the forecast expansion of rubber plantations among countries/province in the wider Mekong Region.

3.5 *Large-Scale Transport Infrastructure*

There are direct and indirect impacts of infrastructure developments. This study assumed the construction of a railway from Kunming to the coast of Cambodia as an example to assess potential land use change triggered by railway construction. According to the Chinese National Standard, railway base width ranges from 4.9 to 6.7 m and a buffering shoulder width of 0.4 to 0.6 m (Chinese Ministry of Railway 2005). A single track railway may occupy a base width ranges from 4 to 4.1 m and two-track railway base width from 14.2 to 18.1 m. The study assumes the widest railway base of 18.1 m to estimate land taken. The six railway projects proposed have a total length of 3,271 km (Mohan-Vientiane Railway Project, no date), requiring 5,920 ha for the railway base. The lack of data precludes the inclusion of operational facilities attached to railway projects. Of the estimated land of 5,920 ha required for railway construction, 4,024.2 ha is forestland, 1,546.8 ha farmland and pasture loss is estimated at 349.0 ha (Table 6.8). The land use change is estimated according to the current land use pattern of Lao PDR.

Assessing implications of large-scale transport infrastructure is complicated. Rowcroft (2008) discusses the impacts of road building on deforestation and reveals complex dynamics: road construction seems to encourage the conversion of forestland into agriculture. In Thailand, forests are fragmented along road and river. Rowcroft (2008) also observes that intensified farming may lead to more profit so that road construction could reduce pressure on forested land in the long run. However, there are other factors: Land use change also depends on economic activities

Table 6.8 The estimated land taken for the proposed railway projects (Units ha)

Total land	Forested land	Farmland	Pasture
5,920	4,024.2	1,546.8	349.0

that occur along transportation corridors. Transportation development can jointly reduce production costs and increase production of goods, productivity and urbanization. Use of farmland will be intensified with increased productivity. Farmland may grow to produce more trading goods, cash crops and fruit trees may be planted. Existing towns along the railway will expand and may need more land for urban development. In Thailand, Vietnam and Yunnan transportation is likely to promote further industrial development, reducing pressure on forest exploitations while Cambodia, Myanmar and Lao PDR are more likely to experience an increased pressure to convert forestland into farmland. On the other hand, experience shows that feeder roads connecting to railway networks are often more important than the railway itself.

3.6 *Large-Scale Mining Activities*

Cambodia, Lao PDR and Vietnam are rich in natural resources especially bauxite. Bauxite mining has the potential to become a large-scale mining operation in all three countries. A concession of 188,095 ha was granted by the Cambodian Government and 97,818 ha by the Government of Lao PDR (Lazarus 2009). A draft mining plan for bauxite was approved by the Vietnamese government in 2007, specifying 180,000 ha in Vietnam's mountainous Central Highlands for bauxite mining. The bauxite mining may amount to 465,913 ha. Mining area may overlap with forested land, coffee plantation and farmland. According to Fox et al. (2009), composition of land use of the Mountainous Southeast Asia, the land taken for mining may manifest in land use change shown in Table 6.9. Forest loss is likely to be the largest land use change followed by the loss of farmland.

Bauxite mining will convert forestland, farmland and pasture into mining operations. The project will acquire land for mining and transportation, processing and urban infrastructure to accommodate mining labour. This will depend on mining operations. Bauxite is found in Bolaven plateau where the majority of Lao's coffee is produced. The total area under coffee production in 2007 was 45,000 ha, representing 30–40% of the total Bolaven land surface area (Galindo and Sallée 2007). Vietnam's Central Highlands are also important production of coffee. It is reported that Vietnam has around 500,000 ha of coffee (Nhan 2010). There is no detailed information on the overlap of mining and coffee production. The land use conflicts may arise between coffee production and mining. It is hard to estimate land use change because mining operations will accelerate the urbanization process and the conversion of farmland to urban infrastructure.

Table 6.9 The estimated loss of different land use types

Land use	% of the total	Area (ha)
Forested land	76.65	357,122
Farmland	13.22	61,594
Pasture	8.69	40,488
Aquatic resource	0.75	3,494
Urban/Built-up	0.69	3,215
Total	100.00	465,913

4 Cumulative Impacts

This assessment concludes that land use change may be significant during 2000 to 2050 considering the influence of the six changes we assumed for the Mekong Region: The expansion of mainstream dams; water diversion; sea level rise and adaptation strategies in Vietnam's Mekong delta; expansion of large-scale rubber plantations; and trans-national transport infrastructure, in particular railway projects linking Kunming, China to Cambodia.

Different drivers cause different change of land use. The expansion of large-scale rubber plantations is likely to result in the loss of native forests, sea level rise in farmland loss and mainstream dams in the loss of wetlands. Mining will compete for land currently used for coffee production and further diminish forests. The area of land converted to transportation infrastructure may not be significant, but there will be substantial indirect effects resulting in land use change.

The possibly accumulated impacts on land use are summarised in Table 6.10. Approximately 2.27 million ha of forest will be converted to commercial plantations and the remainder will be either submerged due to mainstream dams or cleared for transport infrastructure. 850,867 ha of farmland may be converted to 450,070 ha of irrigated farmland while the rest may be lost to sea level rise, mainstream dams and transportation infrastructure. Mainstream dams may submerge 11,966 ha of riverbank gardens, which are unlikely to be re-established due to hydrologic change. The aggregated loss of 78,742 ha of pasture is mainly because of sea level rise and transport infrastructure. 123,029 ha of existing wetlands may be lost to mainstream dams and sea level rise.

The intensive use of farmland will increasingly replace subsistence agricultural practices due to the loss of traditional farmland, increased opportunities for irrigation, improved trans-national transport access and the increasing availability of technology. Expanding demand for agricultural produce may encourage Cambodia, Myanmar and Lao PDR to increase accessible and arable farmland through forest conversion. Improved transportation networks will make farm intensification and expansion more viable as economic activities increase.

All of these drivers may facilitate population concentration or urbanization. Increasing availability of electricity, convenient transportation and irrigation may accelerate the role and extent of agri-business. Access to reliable hydro generated electricity may also promote a process of industrialization. As corollary, industrial

Table 6.10 Accumulated change of land use

Possible decrease	Area (ha.)	Possible increase	Area (ha.)	Note
Forested land	2,269,242	Reservoir	410,118	N.A
Farmland	850,867	Irrigated farmland	450,070	N.A
River bank garden	11,966	See water	476,850	By 2030
Pasture	78,742	Rubber	1,598,000	By 2050
Wetland	123,029	Built-up	5,920	N.A
Built-up	73,025	Mining area	465,913	By 2030
Total	3,406,871	Total	3,406,871	

and manufacturing centres are likely to attract and concentrate mobile labour pools, including households that have lost entitlements to access traditional land and forced to resettle and seek opportunities in urban areas. Towns or cities in the wider Mekong Region may experience rapid expansion as a function of infrastructure development. Land use will shift to urban purposes, especially towns with easy access to mining areas and transportation infrastructure. Landscapes may dramatically change. Natural landscapes may be fragmented by man-made landscapes, such as dams, mines, transportation, urban and recreation facilities.

References

CDRI. 2011. What limits agricultural intensification in Cambodia? The role of emigration, agricultural extension services and credit constraints, *Policy brief no. 7*, 2011.

Chinese Ministry of Railway. 2005. Code for design on subgrade railway, TB10001-2005.

FAO. 1986. *Irrigation water management: Irrigation water needs*. Rome: Food and Agriculture Organization.

FAO. 2004. *From vision to action. A synthesis of experiences in least-developed countries*. Bangkok: Food and Agriculture Organization.

Forest Carbon Asia. No date. http://www.forestcarbonasia.org/in-the-media/LaoPDR-to-increase-forest-coverage.

Fox, J., Y. Fujita, D. Ngidang, N. Peluso, L. Potter, N. Sakuntaladewi, J. Sturgeon, and D. Thomas. 2009. Policies, political-economy, and swidden in Southeast Asia. *Human Ecology* 37(3): 305–322.

Galindo, J., and B. Sallée. 2007. *Participative analysis of coffee supply chain in Lao PDR*. Vientiane: GTC.

Government of China. 2011. http://www.gov.cn/jrzg/2011-12/22/content_2026973.htm.

Hall, D., P. Hirsch, and T. Murray Li. 2011. *Powers of exclusion: Land dilemmas in Southeast Asia*. Singapore: NUS Press.

IWMI. 2006. Water governance in the Mekong region: A need for more informed policy-making. *Water policy briefing*, issue 22, December 2006.

Lazarus, K. 2009. *In search of aluminium: China's role in the Mekong region*. Cambodia: Heinrich Boll Stiftung.

Lu, X. 2007. A study on land policy in GMS. In *GMS study 2007*, ed. J. Qu. Kunming: Yunnan University Press.

Mohan-Vientiane Railway Project. No date. ToR for ESIA.

Molle, F., and P. Floch. 2008. Water, poverty and the governance of megaprojects: The Thai "water grid". *Ambio* 37(3): 199–204.
MRC. 2010. *State of the basin report, 2010*. Vientiane: Mekong River Commission.
MRC. 2011. *SEA for hydropower on the mainstream*. Vientiane: Mekong River Commission.
Nhan, D.T. 2010. Coffee production and trade in Vietnam. *Tea & Coffee Trade Journal*, May 17.
Oguntunde, P.G. 2005. Whole-plant water use and canopy conductance of cassava under limited available soil water and varying evaporative demand. *Plant and Soil* 278: 371–383.
Rowcroft, P. 2008. Frontiers of change: The reasons behind, land-use change in the Mekong basin. *Ambio* 37(3): 213–218.
Ziegler, A.D., J.M. Fox, and J. Xu. 2009. The rubber juggernaut. *Science* 324(5930): 1024–1025. 22 May 2009.
Ziv, G., E. Baran, S. Nam, I. Rodriguez-Iturbe, and S.A. Levin. 2012. Trading-off fish biodiversity, food security, and hydropower in the Mekong river basin. *PNAS* 109(15): 5609–5614.

Chapter 7
Mining in the Mekong Region

Kate Lazarus

1 Methodological Background

This Chapter focuses on an analysis of the mining sector in the Mekong Region. The study in particular explores what development pathways are possible, what are desired and what are likely. It explores the intersection of the mining sector with other pre-determined regional (transboundary) development decisions (although some are not applicable for this sector) as outlined in Chap. 1. This Chapter contains the following sections:

Section 1: Status Quo of the mining sector in the Mekong Region. This Section provides a brief overview of the current state and the recent trends of the mining sector.
Section 2: Probable impacts of prospective changes on the mining sector. This Section synthesises the implications of the major changes/decisions (pre-determined for this study) for the mining sector in the Mekong Region.
Section 3: This Section provides an opinion on the potential cumulative impact if all six changes occurred in the near future on the mining sector.

1.1 Geographic Scope

The focus of this discussion paper is broadly the Mekong Region but will specifically focus on the regional implications of one[1] mining case – bauxite mining in Lao PDR as outlined in Sect. 3.

The Mekong Region (see Figure 1.1) covers the territories, economies, politics and peoples of Burma/Myanmar, Cambodia, China, Lao PDR, Thailand and

[1]During the Expert Panel meeting in January 2011, the team decided to not focus on the Hongsa Lignite mine case.

A. Smajgl and J. Ward, *The Water-Food-Energy Nexus in the Mekong Region: Assessing Development Strategies Considering Cross-Sectoral and Transboundary Impacts*,
DOI 10.1007/978-1-4614-6120-3_7,

Table 7.1 Economic indicators of the GMS (Source: ADB 2010)

GMS economy	Average annual GDP growth (%) 1992–2006	GDP ($ million) 2006	Population (million) 2006	GDP per capita ($) 2006
Cambodia	8.4	7,265	14.2	512
Guangzi	11.7	50,190	49.4	1,016
Yunnan	9.5	60,224	44.7	1,347
Lao PDR	6.5	3,433	5.7	602
Myanmar	9.6	13,002	56.2	231
Thailand	4.5	206,247	65.2	3,163
Vietnam	7.7	60,883	84.2	723
GMS	8.3	401,243	319.6	1,255
China	10.3	2,626,304	1,311	2,003

Vietnam and is home to over 325 million people. Major rivers include the Irrawaddy, the Nu-Salween, the Chao Phraya, the Lancang-Mekong and the Red River. These rivers provide lifelines to numerous peoples whose language, religion and culture have been heavily shaped by the dynamic nature of the region.

The region is also referred to as the wider Mekong Region by the Asian Development Bank. Over the past 30 years, the GMS has seen rapid economic change and is one of the fastest growing sub-regions in the world. The seven GMS economies[2] grew 8.3% on average during 1992–2006 and all economies, except for Thailand, expanded at an average annual rate of at least 6.5% (see Table 7.1). The description of the wider Mekong Region used in this volume approximates the GMS, but does not include Guangxi Zhuang Autonomous Region of China.

2 Status Quo of Mining in the Mekong Region

2.1 Global Mineral Trends Affecting the Mekong Region

Global demand for most commodities has risen dramatically in recent years. Rapid industrial development in China and Thailand as well as in other parts of the world have driven an increase in consumption, especially of fuel and non-fuel minerals. This is coupled with a concurrent increase in consumption of natural resources in developed countries. China accounts for a major proportion of global mineral consumption trends (especially base metals). For some metals (iron ore and copper), Chinese demand growth has single-handedly reversed a trend which would otherwise have led to a decline in global demand for these base metals (World Bank 2011).

[2]The GMS in ADB terms includes the countries of Myanmar, Lao PDR, Cambodia, Vietnam, Thailand and the provinces of Yunnan and Guangxi Zhuang Autonomous Region (referring to '7' economies).

Key noteworthy statistics of relevance include:

- Data for 25 mineral commodities in China indicates that from 1970 to 2004, consumption of many minerals outpaces extraction (hence a need to implement their 'going out' or 'going global' strategy).
- From 2002 to 2005, China accounted for nearly all of the increase in global consumption of lead and zinc. For aluminium, copper and steel, China contributed nearly half of the increase in global consumption. China's total consumption of copper now surpasses that of the US.
- Much of China's minerals consumption can be attributed to infrastructure development and demand for capital and consumer goods; demand for copper, aluminium, and steel will likely increase until at least 2020 (USGS 2004).
- From 2003 to 2006, China's coal exports fell by more than one third while imports more than tripled. Industry sources forecast that China's demand for coal will increase, driven especially by industrialization and the needs of the power sector. China's imports of crude oil are expected to increase dramatically as the number of passenger vehicles per capital grows (USGS 2004).

China has also made substantial investments in metals mining and processing and its own production has grown strongly (Lazarus 2009; World Bank 2011). Chinese mining firms have been strategically looking for opportunities abroad to invest in joint ventures, acquire mining projects and companies, and secure long-term contracts at set price levels. In 2008 China's mergers and acquisitions were worth US$ 52.1 billion, and in the first 2 months of 2009 Chinese companies invested US$ 16.3 billion abroad. China played an important role in providing readily available cash to pull mining companies out of debt around the world (Lazarus 2009). China is also one of the largest producers of many key metals, accounting for a large share of global production of raw steel (35%), lead (31%), zinc (25%) and aluminium (25%) (USGS 2007). However, production of many of these key metals has been hampered by a shortage of concentrates, forcing producers to rely increasingly on imported raw materials, a trend which is expected to continue in the near future.

2.2 *Mining in the Mekong Region*

Mining in the Mekong Region has been small in scale and intensive in labour to date. The vast majority of mine properties are either known deposits with no investors or exploration sites at fairly early (prefeasibility) stages of development (largely in the lower Mekong countries). China accounts for some 80% of recent growth in world demand for minerals. Mining in Lao PDR represents 85% of total exports and contributes to 18% of GDP (Boungoaphalom 2010). The Sepon mine, for example in Lao PDR is a major employer and taxpayer to the country. In the first 2 years of operation, the mine paid the Government of Lao PDR $50 million in tax revenues and $21 million to the provincial government. By 2010, $600 million had been paid to the Government of Lao PDR. Roughly 60% of the company's

3,500 workers come from the local area and the project has contributed to significant infrastructure and efforts to reduce poverty locally. Vietnam has seen dramatic growth in oil and gas and the metals sectors in less than 10 years. Vietnam relies heavily on coal for its energy requirements and there has been a significant rise in coal demand over recent years due to the growing electricity demand and industrial coal requirements, especially from the cement and construction materials industries. The country will continue to exploit and rely on coal-run thermal power plants as the country's hydropower resources have been nearly fully exploited and natural gas resources are limited. Cambodia is the newest entrant in mineral exploration, especially in gold and base metals. Most large-scale mining activities in Cambodia has not reached the exploitation stage however small-scale or artisanal mining is rampant (Oxfam America 2004). As of 2008 there were over 100 exploration licenses granted.

Lao PDR and Cambodia are starting to see a deluge of investors in the mining sector. For example, in Lao PDR along the Bolaven Plateau and in the Central Highlands of Vietnam, China, Australian and international companies investors are starting to consider investing in large tracts of land for bauxite mining to export aluminium, mainly to China, for its growing construction, transportation and packaging industries. BHP Billiton was recently exploring for bauxite in Cambodia's Mondolkiri province but in early 2009 pulled out of the country. Bauxite still exists and it is expected that other investors will move in to carryout their own exploration.

The Mekong Region possesses substantial mineral resources, including significant deposits of gold, bauxite/alumina, cobalt, lead, copper, iron ore, zinc and industrial minerals (limestone, gypsum, potash and clay). Gold, copper, bauxite, iron ore, and industrial minerals are present in Cambodia, Lao PDR and Vietnam. Vietnam also has deposits of anthracite coal, antimony, rare earths, silica, silver, gemstones, tin, titanium and zinc. Thailand has significant deposits of industrial minerals, and less abundant metal deposits. Burma's most important deposits consist of precious and semi-precious gemstones, natural gas, and copper. Table 7.2 presents examples of mineral commodity trade flows of the Mekong Countries.

2.3 *Institutional Drivers*

2.3.1 Actors

Most large-scale mining activities in the Mekong Region are led by companies from China, Australia and Canada. The five largest companies with interest in the region are BHP Billiton, Alcoa, Alcan, Ivanhoe Mines and MinMetals. There are numerous other small-scale mining countries from China, Thailand, Vietnam, Lao PDR, Korea, etc. with limited information in the public domain.

Table 7.2 Mineral commodity trade flows of wider Mekong region countries (2005–2006)

Country	Commodity	Net trade balance	Primary trading partners
China	Copper, tin, manganese, iron ore, sand, gravel	Negative	Myanmar, Lao PDR, Vietnam, Thailand
China	Gravel, sand	Positive	Lao PDR, Thailand
Lao PDR	Sand, tin	Negative	China, Thailand
Lao PDR	Gravel	Positive	Thailand
Myanmar	Copper, tin	Negative	China, Thailand
Myanmar	Iron ore, manganese, gravel	Positive	China, Thailand
Thailand	Copper	Negative	Chile
Thailand	Iron ore, manganese, gravel, sand	Positive	China, Lao PDR, Cambodia, Malaysia
Vietnam	Aluminium, manganese, gravel, sand	Negative	China, Thailand, France (gravel), Australia (sand)
Vietnam	Copper, aluminium, limestone, sand	Positive	China, Japan, Rep of Korea

Source: UN Comtrade (2007)

Note: Import/export data are incomplete and often under-reported. Data for Cambodia was not available. When the trade balance is negative, the country is a net importer; a positive trade balance indicates the country is a net exporter

Financial institutions have limited involvement in this sector in the Mekong Region. This is partly due to the fact that most activities are still in exploratory phases. Some banks associated with mining projects in the region include Standard Bank of London, HSBC, ANZ Bank, WestLB.

The World Bank has been significantly involved in the mining sector in Lao PDR. Their involvement first began with IFC involvement in Oxiana's planned Sepon gold mine. The World Bank did not become a financier but used the opportunity to engage the government in modernising its mining law, taxation regime and developing its internal management capacity. In 2006, the World Bank commissioned a number of studies related to Lao's mining sector and between 2007 and 2011 supported the revision of the Mining Law. In 2011, the World Bank released the "Lao PDR Development Report 2010 Natural Resource Management for Sustainable Development: Hydropower and Mining" (Molle and Floch 2008). In addition, the World Bank has a Technical Assistance Grant to the Government of Lao PDR for capacity building in the mining sector (2010–2014) with support from AusAID.

The Asian Development Bank has been primarily involved in infrastructure development with some attention provided for hydrocarbon investments and pipelines. In Cambodia, the ADB has been promoting the development of mid-stream and downstream aspects of the hydrocarbons sector.

AusAID, NORAD and JICA have been involved in providing technical assistance, promoting tax and legislative reform, advising on contract negotiations and developing the regulatory capacity in Cambodia's hydrocarbon sector. The Bank of China is heavily involved in promoting investments of Chinese companies.

2.3.2 Institutional Challenges

There are many institutional challenges in the region:

- Limited capacity in geological information and geological exploration to examine mines for mineral resources in most of the countries but namely Lao PDR and Cambodia (this is particularly significant as the mining sector is expected to grow in these two countries)
- Private sector driven development: No competitive bidding process for specific sites
- Risks in investment – particular concerns of uncertainty about legal and fiscal terms
- Most mineral laws are outdated or are currently under review thus requiring timely promulgation of Minerals Law in order to address the rapid investment climate, although Lao PDR as recently updated its law and is expecting to call a halt to future investments
- The enabling environment for private investment is still weak in most of the countries. For example:
  - Lack of protection for investors
  - Ease of doing business low in Cambodia and Lao PDR
  - Lack of confirmation of security of tenure and 'finder's rights to mine'
  - Lack of confirmation of competitive fiscal terms to attract investors ('fair' benefit-sharing between investors and government)
  - Limited level playing field to treat all investors equally; most countries have two decision-making processes in place.

3 Analytical Approach and Indicators

The World Bank (2011) developed three scenarios of which to base projections around: (1) low, (2) base and (3) high cases of global economic activity and derived impacts in the mining sector. Here I try to use them as indicators to determine change and impact around the key regional decisions determined by the EMRF project.

Table 7.3 Assumptions based on low, base, high scenarios

Price assumptions (low case (−10%)/high case (+20%))			
Commodity	**Reference**		
Alumina	$250 US$/t		
Production assumptions			
Alumina	**2009–14**	**2015–19**	**2020–24**
Low case (tons)	n/a	n/a	n/a
Base case (tons)	n/a	1,000,000	1,000,000
High case (tons)	n/a	1,000,000	1,000,000
Cost assumptions			
Alumina refining	Profit tax assumption based on feasibility study figures		

Source: Adapted from World Bank (2011)

Table 7.4 Base case annual production

Indicator	Alumina
Production/export value US$ p.a. (17 year average)	2015: 250 million
Production volume p.a.	2015: 1 million tons
Fiscal receipts p .a. (including dividends) US$ (17 year average)	5.3 million
National jobs	600–1,000

Source: Adapted from World Bank (2011)

The assumptions for mining made by the World Bank particularly relate to the decisions around bauxite which these indicators may be particularly useful (Table 7.3).

Mining operations in Lao PDR were severely impacted by the economic downturn with all expansion plans being deferred as a consequence of the credit crisis. During that time, operations with high operational costs were stopped in response to low commodity prices. The low case scenario is predominantly focused on negative price fluctuations rather than reduction in production figures (World Bank 2011) however alumina is not affected by this given there is no current exploitation of the resource.

The base case scenario is founded on business plans for existing operations as expected until the financial turbulence erupted at the global scale and metal prices plunged in the second half of 2008. This scenario expects that an alumina refinery, which is currently under review, will become operational by 2015 (Table 7.4).

The high case scenario assumes that metal prices will stabilize at a level 20% higher than forecasts. This implies a 'price plateau', which is 50–100% higher than the trend level of 1985–2005 but still markedly below the 'super-cycle' peak level, which was attained during 2007–2008. It is assumed that all operations will resume to their full expansion programs and small and medium scale operations will develop after 2015. Alumina is one such mineral that has shown good potential and is expected to come into commercial operation within 5–10 years. It is common practice that a medium/large-scale mine operation will take at least 7 years to develop: around 4 years of comprehensive exploration, followed by feasibility study and mine development which takes around 18 months.

Production: Estimates on production of the industry are important when discussing the purpose and need for a mining sector to develop further. Projections of realised mine potential are best defined in terms of **annual production volume and capacity**, and not in financial terms since price volatilities can mask other more fundamental trends. For example, in Lao PDR, the combined export value for 2007 was at a level close to 2008 despite a 50% increase in copper production and 25% increase in gold production in 2008, the two dominant commodities in the country. Even an additional 33% increase in gold and copper volumes which were planned for 2009 are not likely to compensate for the low metal prices which were sustained during the first quarter of 2009 (World Bank 2011).

Project Volume: Projects based on the number of mines are difficult to covert to volumes of production, because the size of a commercially viable mine may vary tenfold within the specific commodity and even more across commodities.

Levels of investment/Fiscal Receipts: This is also a useful indicator for those interested in the types of companies or entities involved in mining in the region as well as associated decision-making practices. However, pledged levels of investment (or even approved investment) does not often show the true reality of investment in a project due to various factors such as length of time for approval of environmental and social documents, decision-making factors of host governments, company's own decision-making process, reality of the availability of the resource after exploration, etc.

Confirmation of available resources: This indicator can also be useful in assessing impacts to the mining sector however it can also be misleading because in some of the countries there are modest exploration activities and the knowledge required to confirm resource availability is limited. This is not to say that the resource potential is weak. Quality of the mineral resource is also important but again difficult to assess. However, it this stage it can give us an indication of the potential impact on the sector based on known available resources.

National jobs: Creation of jobs is a useful indicator to measure as to a key reason why the government may engage in this particular development strategy. However, experience from Lao PDR shows that these are often highly technical positions and whilst some companies make efforts to employ local people, many jobs are taken by migrants from neighbouring countries or from developing countries or local elite.

The following sections discuss changes by some of the decisions in isolation as well as their cumulative impacts.

3.1 Mainstream Dams

After discussion at the partners meeting in January 2011, it was agreed that the impacts of the decision to go forward with mainstream dams would not have significant impacts with regards to electricity issues for the mining sector. However, sand mining which is a major exploitation activity in the Mekong delta (see 4.2), will likely be significantly impacted by a decrease in levels of sand if the mainstream Mekong dams go forward (see WUP-FIN documents).

3.2 Sea-Level Rise and Adaptation Strategies in Vietnam's Mekong Delta

The Mekong delta accounts for most of Vietnam's sand exports. In 2009, Vietnam shipped 9.2 million cubic meters of sand (seven times more than in previous years). There are more than 150 licenses, which allow around 321.6 million cubic meters of sand to be minded in the delta. Intense sand mining activities in the delta channels as well as the construction of reservoirs in the upper reaches of the Mekong River could be envisaged as direct man made contributions to the deterioration of the Mekong delta. In combination with the ongoing sea- level rise substantial

Table 7.5 Summary of impacts from sea-level rise

Indicators	Impact
Production	Over-exploitation of sand mining in Vietnam delta expected as reported in 2009
Volume of resource	Volume expected to increase as need increases
Level of investment/fiscal receipts	
Availability of resources	Expected to decrease
Employment	Overtime will increase as production increases but would expect to decrease by 2030 when resource is depleted

land-loss of more than 30% in the densely populated delta plain during this century must be expected (Stattegger et al. 2010) (Table 7.5).

3.3 Rubber Plantations

Natural rubber has been promoted as a major export crop in Thailand and, more recently, in Vietnam where growing areas have expanded rapidly since the early 1980s in response to high global market demand and promotion by the governments. In contrast, efforts to expand rubber cultivation are a relatively new trend in Cambodia and Lao PDR, mainly driven by overseas investors (Baumuller 2008). More than 500,000 ha have been converted already in the upland areas of the Mekong Region (Ziegler et al. 2009). It is expected that land converted to rubber will double or triple by 2050. Over 20 reports have been conducted on rubber alone in Lao PDR, noting the significant impacts on the environment and people.

Potential effects on the mining sector include:

- Competition is increasing among investors and government agencies as large-scale plantations for rubber compete with other land uses such as high value coffee plantations, mining and hydropower.
- Large-scale plantations contribute to degraded slopes thus impacting fragile soil that could be important for mining activities.
- In several countries in the Mekong (i.e. Lao PDR and Cambodia), allocation of concessions is often overlapping among various sectors including hydropower, mining and land. There is often little knowledge of which concessions are located where leading to mismanagement and conflict among differing actors. Further, there is a gap in the regulatory framework in many countries concerning logging in non-productive forests in the case of concessions for mining (and hydropower development).
- In one example from Lao PDR, where concessions allocated by the army are commonplace, a rubber concession contract also gave away mining rights and other types of resource claims within the concession range to the company (typically if additional resources are discovered, the excavation rights are to remain with the Lao Government (Shi 2008) (Table 7.6)).

Table 7.6 Summary of impacts from rubber plantations

Indicators	Impact
Production	Production expected to increase over medium term due to expansion and development of existing projects. Some projects currently under survey will switch to commercial operation. Over longer term, production will decrease as Government of Lao has made a new decision to halt new mining investments and promote agriculture as top priority for investment and development. Rubber plantations are one such agricultural commodity which will be expanded
Volume of resource	Volume of resource may not change but production will decrease (as noted above)
Level of investment/ fiscal receipts	Mining products vary considerably in price with some commodities susceptible to volatile market forces and the world economic situation. With the expansion of rubber plantations and the competition of concessions, levels of investment may decrease
Availability of resources	Decrease over time
Employment	Job creation will be slower due to the technical skills required and agriculture production promotion

3.4 *Large-Scale Transport Infrastructure Investment*

3.4.1 Connecting the Mekong Region by Railways

Connectivity and improving transport in the Mekong Region is one of the cornerstone activities of the ADB's GMS Economic Cooperation Program. In particular, railways were identified as a key priority for development. These plans have been recently accelerated due to escalating sub-regional trade, growing concerns over climate change, and more recently, sharply fluctuating fuel costs (ADB 2010) (See Figure 1.6). Cambodia's railway is being rehabilitated by the ADB and AusAID while China is financing a feasibility study for the line between Phnom Penh, Cambodia to the Vietnam border. China is building a new line to the Vietnam and Myanmar borders. Thailand is considering developing high-speed train lines and new lines to link to Lao PDR and onward to Vietnam. Thailand is also considering a policy shift from road to rail. Vietnam is also considering lines to Lao PDR and Cambodia.

3.4.2 Railway from China to Lao PDR

The Lao and Chinese Governments recently signed an agreement to construct a 421 km railway from the China-Lao border in Luang NamTha Province at Boten (the site of a Chinese run casino) to the Lao-Thailand border near Vientiane (See Figure 1.6). The project is expected to cost US$7 billion and construction is expected to start on 25 April 2011 to mark the 50th anniversary of Lao PDR-China diplomatic relations (Vientiane Times 2010b) and be finished in 4 years. The Chinese government will hold a 70% stake in the project with the remaining held by the Lao Government. The railway is expected to be part of the Asean-China railway, which

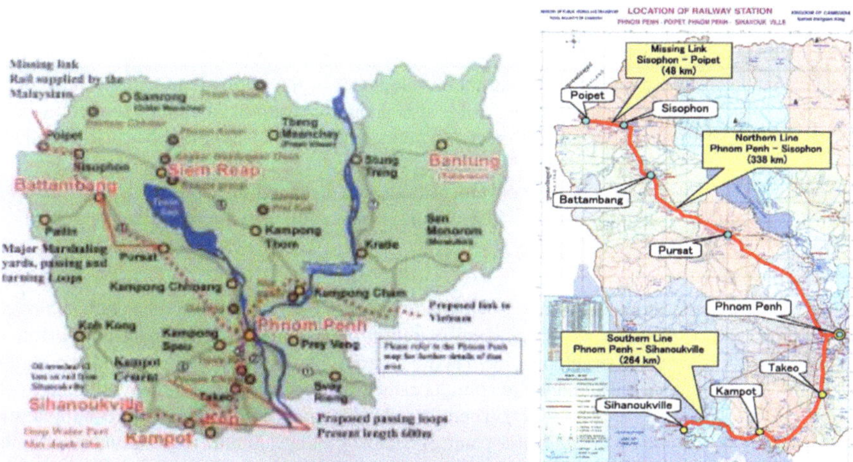

Fig. 7.1 Railway plans in Cambodia (Source: Toll Royal Railways 2010, http://www.tollroyalrailway.com/)

runs from Yunnan Province south through Lao PDR to Thailand, Malaysia and Singapore. According to the Vientiane Times (2010b), "the project is a step towards China's goal of increasing trade value with ASEAN Nations after the free trade agreement between the two parties came into effect at the start of 2010". The Thai government is partnering with China to improve its rail link from the Lao-Thai border in Nong Khai to Bangkok. The Government of Lao PDR (2010) plans to provide land as its share in the venture while the Chinese partner (along with the Chinese government) will provide the funding for the construction of the railway (www.laopdr.gov.la). The railway is expected to accommodate medium speed electric trains which can run at about 200 km/h. The highest speed trains will run at 350 km/h.

3.4.3 Railway in Cambodia

The Royal Government of Cambodia has outsourced its railway operations under a 30 – year exclusive concession to Toll (Cambodia) Co., Ltd (Trading as Toll Royal Railway) to operate the Cambodia Railway Network (NewsDesk 2010). The project is funded by the Asian Development Bank (ADB) and AusAID for US$140 million for the network and infrastructure upgrade. The upgrades (see Figure 1.3) include the South Line (254 km) from Phnom Penh to Sihanoukville, the North Line (388 km) from Phnom Penh to Poipet (line rehabilitation) and the Missing Link (48 km) from Poipet to Sisophon (line reinstatement) Fig. 7.1.

3.4.4 Likely Impacts of Railway Infrastructure on the Mining Sector

A study in 2009 concluded that all mining companies would be looking for the cheapest opportunities to transport mineral resources for export, mostly to China

Table 7.7 Summary of impacts from railway infrastructure on the mining sector

Indicators	Impact
Production	China is facilitating the development of the railroads in many of the countries in the Mekong and is also a major investor in the mining sector. It would be expected that greater benefits would go to China first because the goods exported (from Lao PDR and Cambodia to China) are expected to increase sharply due to convenient transport. It would be expected that the mining sector in both countries would significantly increase thus creating a 'rush' to invest in mining
Volume of resource	
Level of investment/fiscal receipts	Bauxite mining is emerging in Lao PDR and Cambodia with expected exports to China. Transportation of bauxite is a significant consideration when developing the bauxite industry. Few countries in the world export raw bauxite, and while alumina can be transported, it is extremely difficult to do so (Lazarus 2009). Therefore, the countries are looking at options for developing the full value chain of bauxite mining, alumina refining and aluminium smelting. Currently there are transport links between Cambodia and Lao PDR to Vietnam. However, it was thought that from Vietnam the product would travel to China. With the new railway links in place, it is probable that either alumina or aluminium will be transported via railroad to China creating a direct connection between the mining source and the market. In another case, potash mining in Lao PDR which is also seeing numerous investors come into the country is expected to be exported 100% to overseas markets (Vientiane Times 2010a). The main market is China, while some will be exported to Thailand. The company, Sinohydro, expects that the railway (of which the company is also developing) will open up more market opportunities for the potash mining and factories (Vientiane Times 2010a)
Availability of resources	
Employment	Labour issues in the mining sector will also be significantly impacted as mining will sharply increase so will more foreign labourers. In the case of Chinese companies, they often bring their own labourers who do not leave. The Lao Government indicates that projects are designed to partly provide employment to the domestic population but often is the case that there are minimal Lao employees (partly due to capacity issues)

(Lazarus 2009). Are the new railroad lines being built in Lao PDR and Cambodia the answer to exporting mineral resources? Possible impacts from the railroad infrastructure on mining (Table 7.7).:

3.5 *Bauxite Mining in Lao PDR*

The triangle area of northeastern Cambodia, southern Lao PDR and the Central Highlands of Vietnam is a strategically important area for cross-border bauxite mining (see Figure 7.1). Cambodia's Mondulkiri province in the northeast of the country is one of the poorest and it is located 521 km northeast of Phnom Penh.

The main livelihood activities of the province include farming, fishing and non-timber forest products and increasingly large-scale agriculture production and mineral exploration are putting pressure on local livelihood resources. The Bolaven plateau in southern Lao PDR is located in the northeast of Champassak province and spreads into Salavan, Sekong and Attapeu provinces. The plateau is of volcanic origin, has an average elevation of 1,100 m, and its soil is composed of fertile red basaltic soil. Ninety nine percent of Lao PDR' coffee is produced on the Bolaven plateau. The vast majority of Vietnam's bauxite mines are located in the Central Highlands of which Dak Nong province contains 60% alone. The Central Highlands is a stunningly beautiful and fertile region of Vietnam, and fertile forested mountains, coffee plantations and an enormous ecotourism potential characterise the region.

3.5.1 Background

Deposits of bauxite have been prospected and explored for the last 4 years in the Bolaven plateau in southern Lao PDR. One firm has received a production agreement and seven others are still prospecting. A production agreement allows the firm concerned to start the mining operations. The size of the deposit concerned is estimated at 300 million tons of exploitable bauxite. The firm is considering investing in alumina refining on the basis of 500,000 t of alumina output. It has started to investigate the commercial potential at prefeasibility level. Alumina is produced by a chemical process, the Bayer process. Its most significant use, 85% of the world alumina production, is in the production of aluminium metal. The remainder is used as an abrasive due to its hardness and as a refractory material in a wide variety of processes due to its high melting point. At a later stage, the firm concerned may consider investing in an aluminium smelter, possibly in consortium with other investors.

Dependent on the technology to be applied and the interests and 'configuration' of the potential investor, a commercially viable alumina refinery would require upwards of 0.5 million tons of alumina output or between 1.0 and 1.5 million tons of raw bauxite per year, dependent on the grade of the bauxite, for a period of not less than 20 years. Such a venture would need a considerable volume of electricity, possibly in the region of 150 MW and large quantities of water.

The environmental implications of opencast bauxite mining cum alumina refining are vast. A large concession area would be required and would have to be fully cleared. The wastewater from the refinery is highly polluted. The waste is called 'red mud' which is difficult to process or use for other purposes. Further downstream industrial development would be processing of the alumina in an aluminium smelting operation into saleable aluminium, which is a relatively clean process and requires 100% reliable electricity in the order of 600–800 MW, dependent on the technology applied and the scale of the operation. The full investment required for alumina refining, aluminium smelting and hydropower development might run into US$ 5 billion.

Decisions around whether or not to move forward with bauxite mining in Lao PDR are complex. Historical government practice in the hydropower and mining sector has been one of developers proposing projects to the GOL rather than conforming to model documentation and standards set by GOL. In 2009, the GOL requested the World Bank to provide them with global information about bauxite mining given the fact that no one in the government had any familiarity with bauxite mining nor the environment or social impacts and fiscal benefits. If bauxite mining takes place in the Bolaven plateau, other industries will be significantly impacted (such as tourism and agriculture). As a sector, mining in Lao PDR is one of learning by doing and driven by private sector developers.

In order for bauxite mining to go forward, the government needs to have the institutional and regulatory standards for mining properly in place. In 1997 the Lao Mining Law was approved, followed by an Implementing Decree in 2002 to try and facilitate development in this sector through FDI. The mining sector is considered one of the top sectors for generating income and reducing poverty. Over the years there has been uncontrolled granting of a number of mining concessions at the provincial level and poor environmental monitoring. Part of this results from the lack of knowledge and enforcement of the law at the local levels coupled with capacity gaps. In conjunction, the Foreign Investment Law, which includes the levels of investments for approval by central or provincial authorities is not well known or understood. Similarly, the institutional mechanisms have been constantly updated and changed (e.g. from the Committee for Planning and Investment to the Ministry level to roles and responsibilities). In 1996, a moratorium on mining was put in place. The main reason behind the moratorium was the administrative burden associated with negotiating complex agreements on a project-by-project basis, making it difficult for the Ministry of Energy and Mines (MEM) to regulate the sector (Rutherford et al. 2008). In 2007, the government announced a slow down/moratorium on granting new exploration and production licenses In order for the government to carryout of existing practices concerning taxation, operational requirements and other contractual obligations (Rutherford et al. 2008). The government also planned to cancel agreements with non-performing companies. Over the last 2 years, the GOL has been developing a new Minerals Law which they suggest will help to centralise and control the granting of mining concessions for large projects more effectively and facilitate greater FDI into Lao PDR (Table 7.8).

4 Cumulative Assessment

What would the combined impact on the mining sector if all these changes occurred in the near future?

I believe that all the decisions reviewed in Sect. 4 will eventually go forward but the timing of them is very uncertain and the configuration of agreements is unlikely to be known. These decisions will be based, I believe, on the drive of the private sector to invest as well as powerful government actors, such as the Chinese government which require certain resources to sustain their economies.

Table 7.8 Summary of impacts from bauxite on the mining sector

Indicators	Impact
Production	Export revenues are expected to increase based on the World Bank's high case scenario
Volume of resource	
Level of investment/fiscal receipts	Fiscal impact would be limited due to tax holiday's covering the first 7 years of operation i.e. beyond 2020. After transferring to full tax obligations, the combined fiscal receipts from alumina refining are not expected to exceed $US10 million/year because of preferential tax status and world market prices below past years' historical highs, and possibly transfer pricing below actual market price, as suggested in the current mine/refining configuration
	Alumina and aluminium prices saw drops by 70% during the second half of 2008 before moderate recoveries in the first half of 2009
Availability of resources	Overall decrease from rubber plantation expansion
	Overtime the resource will be depleted. Reports indicate availability until 2030 at least. What is left behind is often environmental disasters from red mud ponds and other chemical inputs
Employment	Job creation will be slower due to the technical skills required and agriculture production promotion
	Local people will be sourced but as this is highly technical, it is envisaged that not many jobs will be provided to the local Lao

Based on the discussion in Sect. 4, Table 7.9 outlines the effects from each of the decisions with a view towards determining their cumulative impacts.

Based on Table 7.9, the mining sector benefits significantly from some of the regional decisions expected. This win translates into increases in GDP for many countries in the region – most notably Lao PDR where we discuss the bauxite and lignite mining industry.

With up to seven investors reportedly exploring for bauxite, political interest in the development is strong however a number of uncertainties still exist including the supply of the resource, the rights to the resources (in terms of who has concessions already and who does not) and the environmental and social impacts. The transportation linkages provided by the new railways facilitates a new avenue for transporting alumina paving the way for alumina refining in the country. Smelting of aluminium is also being considered but it requires large amounts of cost-effective electricity of which may be available from mainstream Mekong dams.

An increase in seawater in the Mekong delta could significantly hamper existing mining operations.

Overlapping concessions for hydropower, mining and plantations are continuing to cause conflict and show the lack of planning for such investments. Corporate investors will win big with the investments however the Government of Lao PDR may benefit in the long-term (not the short term given the length of time for mining projects to commence) and local people will be significantly impacted economically, environmentally and socially.

Table 7.9 Cumulative development effects on the mining sector in the Mekong region

Indicators	Impact
Production	Related to sea-level rise, expected over-exploitation of sand mining in Vietnam delta as reported in 2009
	Production of mining in Lao PDR expected to increase over medium term due to expansion and development of existing projects. Some projects currently under survey will switch to commercial operation. Over longer term, production will decrease as Government of Lao has made a new decision to halt new mining investments and promote agriculture as top priority for investment and development. Rubber plantations are one such agricultural commodity which will be expanded
	China is facilitating the development of the railroads in many of the countries in the Mekong and is also a major investor in the mining sector. It would be expected that greater benefits would go to China first because the goods exported (from Lao PDR and Cambodia to China) are expected to increase sharply due to convenient transport. It would be expected that the mining sector in both countries would significantly increase thus creating a 'rush' to invest in mining
	Export revenues are expected to increase based on the World Bank's high case scenario
Volume of resource	Volume of sand for mining expected to increase as need increases
	Volume of resource in Lao PDR may not change as a result of rubber expansion but production will decrease
Level of investment/ fiscal receipts	Mining products vary considerably in price with some commodities susceptible to volatile market forces and the world economic situation. With the expansion of rubber plantations and the competition of concessions, levels of investment may decrease
	Bauxite mining is emerging in Lao PDR and Cambodia with expected exports to China. Transportation of bauxite is a significant consideration when developing the bauxite industry. Few countries in the world export raw bauxite, and while alumina can be transported, it is extremely difficult to do so (Lazarus 2009). Therefore, the countries are looking at options for developing the full value chain of bauxite mining, alumina refining and aluminium smelting. Currently there are transport links between Cambodia and Lao PDR to Vietnam. However, it was thought that from Vietnam the product would travel to China. With the new railway links in place, it is probable that either alumina or aluminium will be transported via railroad to China creating a direct connection between the mining source and the market. In another case, potash mining in Lao PDR which is also seeing numerous investors come into the country is expected to be exported 100% to overseas markets (Vientiane Times 2010a). The main market is China, while some will be exported to Thailand. The company, Sinohydro, expects that the railway (of which the company is also developing) will open up more market opportunities for the potash mining and factories (Vientiane Times 2010a)
	Fiscal impact would be limited due to tax holiday's covering the first 7 years of operation i.e. beyond 2020. After transferring to full tax obligations, the combined fiscal receipts from alumina refining are not expected to exceed $US10 million/year because of preferential tax status and world-market prices below past years' historical highs, and possibly transfer pricing below actual market price, as suggested in the current mine/refining configuration
	Alumina and Aluminium prices saw drops by 70% during the second half of 2008 before moderate recoveries in the first half of 2009

(continued)

Table 7.9 (continued)

Indicators	Impact
Availability of resources	Availability of sand in the Vietnam delta expected to decrease
	Availability of mineral resources in Lao PDR will decrease overtime as concessions are provided to agricultural development rather than mining
	Overtime bauxite will be depleted. Reports indicate availability until 2030 at least. What will be left behind is often environmental disasters from red mud ponds and other chemical inputs
Employment	Overtime will increase as sand mining increases but would expect to decrease by 2030 when resource is depleted
	Labour issues in the mining sector will also be significantly impacted as mining will sharply increase so will more foreign labourers. In the case of Chinese companies, they often bring their own labourers who do not leave. The Lao Government indicates that projects are designed to partly provide employment to the domestic population but often is the case that there are minimal Lao employees (partly due to capacity issues)
	Job creation will be slower due to the technical skills required and agricultural production promotion
	Local people will be sourced but as this is highly technical, it is envisaged that not many jobs will be provided to the local Lao

5 Conclusion

There was a significant gap in information on mining in several of the countries, especially latest mining statistics and many of the decisions reviewed lack available information.

However, this Chapter aims to get a sense of the mining sector in the Mekong Region and how the sector will be changed or impacted by four potential development decisions namely: sea level rise and adaptation strategies, rubber plantations, railways and bauxite mining.

If all of the projects take place there is likely to be an increase in the mining sector in the Mekong Region as a major contributor to GDP and in the name of poverty reduction. Doors will be opened wider for investors to channel their funding as exploitations becomes more accessible energy may be more widely available and transportation linkages provide direct routes for export.

Uncertainty comes with all of these decisions, as institutional and regulatory systems are not currently in place in some of the countries to facilitate the influx of mining activities.

However we cannot assume that all of these project decisions will line up in terms of timing given that mining and hydropower projects take a long time to reach completion. For example, whilst bauxite mining may come online in the next year or two, the cheap energy sources to facilitate the refining and processing may not be available for 5–10 years.

Finally, while the railway transport is moving 'full steam' ahead, one question remains as to who are the passengers?

References

ADB. 2010. *Connecting greater Mekong subregion railways: A strategic framework*. Manila: The Bank.

Baumüller, H. 2008. Prospects and drivers for agricultural change in the Mekong region: The case of rubber, rice and sugar. Vientiane: WWF Greater Mekong Programme.

Boungoaphalom, E. 2010. Status of mining in Lao PDR. Presentation at the international conference on mining: Staking a claim in Cambodia, 26–27 May 2010.

Government of Lao PDR. 2010. China to cooperate in railway construction. www.laopdr.gov.la. Website. Lao PDR, Accessed 14 Dec 2010.

Kummu, M. 2008. Spatio-temporal scales of hydrological impact assessment in large river basins: The Mekong case. Doctoral dissertation. Helsinki: Helsinki University of Technology.

Lazarus, K. 2009. In search of aluminium: China's role in the Mekong region. Geneva: Heinrich Bolle Stifling, WWF and International Institute for Sustainable Development.

Molle, F., and P. Floch. 2008. Megaprojets and social and environmental changes: The case of the Thai "water grid". *Ambio* 37(3): 199–204.

NewsDesk. 2010. Toll royal railway opens phase one of the Cambodian railway. 25 October. http://khmerization.blogspot.com/2010/10/toll-royal-railway-opens-phase-one-of.html.

Oxfam America. 2004. *Small-scale gold mining in Cambodia: A situation assessment*. Phnom Penh: Oxfam America.

Rutherford, J., K. Lazarus, and S. Kelley. 2008. Rethinking investments in natural resources: The role of China in the Mekong region. Geneva: Heinrich Bolle Stifling, WWF and International Institute for Sustainable Development.

Shi, W. 2008. Rubber boom in Luang Namtha: A transnational perspective. GTZ.

Stattegger, K., D. Unverrricht, and R. Tjallingii. 2010. *Evolution of the Mekong delta: Holocene history, present situation and perspectives for the future*. Kiel: Institute of Geosciences, Christian-Albrechts-University.

Toll Royal Railways. 2010. Maps of the railway linkages in Cambodia. www.tollroyalrailway.com.

UN Comtrade. 2007. Mineral commodity flows data.

USGS. 2004. Data for mineral imports for China. USGS Minerals Commodities Surveys. minerals.usgs.gov/minerals accessed 10th February 2013.

USGS. 2007. Data for mineral imports for China. USGS Minerals Commodities Surveys. minerals.usgs.gov/minerals accessed 10th February 2013.

Vientiane Times. 2010a. Chinese firm builds potash processing factory. December 24.

Vientiane Times. 2010b. Government accelerates Lao PDR-China railway preparations. December 25.

World Bank. 2010. Lao PDR development report 2010: Natural resources management for sustainable development, background paper on economic assessment of the future of the Lao mining sector.

Ziegler, A.D., J.M. Fox, and J. Xu. 2009. The rubber juggernaut. *Science* 324: 1024–1025.

Chapter 8
Cross-Sectoral Assessment

Alex Smajgl and John Ward

1 Methodology

Accounting for disciplinary frontiers and methodological impediments presents a formidable challenge in assessing the outcomes of single, independent development decisions in a sustainability context. Assessing the outcome of multiple, contemporaneous large-scale development investments requires innovative methods and processes. As part of a larger project, the *Exploring Mekong Region Futures* project conducted the expert panel assessment to explore cross sectoral and trans-boundary trade-offs of individual development strategies and how cumulative effects might evolve. Theoretically, such an assessment could identify multiple strategies characterised by:

- Uni-directional positive effects,
- Uni-directional negative effects, or
- Mixed sign effects with some effects balancing out other effects.

Effects of mixed sign are the most likely category as the same investment can have a positive connotation for one sector and a negative impact on another sector. Synthesising mixed effects as assessed by the six sectoral assessments in the previous Chapters into one composite metric would render all relevant information into neutralised statements that convey minimal insights. As an alternate strategy, this final Chapter describes a cross-sectoral synthesis for each of the six identified development strategies and, in conclusion, proposes a cumulative perspective. Table 8.1 summarises the approach.

Development strategies have typically one primary goal defined by one particular sector. For instance, the primary goal of hydropower is the generation of electricity. Consequentially, the assessment of a specific investment is likely to appear more balanced or positive within this sectoral perspective. In contrast, assessments conducted through the lens of other sectors are likely to appear much less balanced, in particular due to flow-on effects that eventuate with higher levels of uncertainties. Therefore, some of the previous Chapters have described a sector specific presentation, potentially

A. Smajgl and J. Ward, *The Water-Food-Energy Nexus in the Mekong Region: Assessing Development Strategies Considering Cross-Sectoral and Transboundary Impacts*,
DOI 10.1007/978-1-4614-6120-3_8, © Springer Science+Business Media New York 2013

Table 8.1 Sectoral and cross-sectoral assessment perspectives

		Six impending investment decisions (columns : ⇩ Chap. 8)						
		Hydro	Water diversion	SLR	Rubber	Mining	Railway	Cumulative
Six sectoral dimensions (Rows): Chaps. (2, 3, 4, 5, 6 and 7) ⇨	Water							
	Food							
	Energy							
	Livelihoods							
	LUC							
	Mining							

LUC represents land use change and SLR represents sea level rise

polarising reader interpretations. The polarisation of values and beliefs resembles decision-making processes, whereby competing value orientations are mobilised, confined and perpetuated within the boundaries of agencies institutionalised to represent prescribed sectoral interests, such as securing required energy supplies.

The Chapter intentionally emphasises the sectoral benefits and/or costs accruing to water, food and energy and synthesises sector specific debates occurring in the wider Mekong Region. Aggregating independently evaluated sectoral effects coupled with an elaboration of proposed interactions enable a more precisely focussed appraisal of the actual trade-offs of individual development strategies. We propose that this a central proposition of investigations of the water-food-energy nexus. The next sections seek to provide the foundation of a more comprehensive and tractable nexus perspective. To avoid repetition we focus on key sectoral consequences and the processes by which ripple effects might eventuate.

2 Hydropower

The development of mainstream hydropower projects in the wider Mekong Region substantially affects the water-food-energy nexus. Energy security is most likely improved, water security will be substantially redistributed and the cumulative impacts on food security are uncertain. The key influences identified are

- The increase of power generation capacity (although the relative merits of this increase would not be as substantial if compared to the expected increases in energy demand);
- The temporal and spatial redistribution of capacities to utilise water;
- The provision of irrigation potential; and
- The loss of agricultural land due to inundation and accelerated industrialisation and urbanisation.

These key influences are triggered by a set of interrelated impacts described in Chaps. (2, 3, 4, 5, 6 and 7 of this Volume.

Hydrological impacts include the substantial increase in storage capacity and dry season inundation, increasing dry season flow, delayed flood timing, reduced flooded area, and reduced sediment transportation (Chap. 2). Importantly, the magnitude of positive impacts such as increased dry season flow are predicted to be offset by reduced snow melt and to diminish further downstream. Most positive impacts remain in Lao PDR, for instance the substantial increase in water storage and irrigation potential. Cambodian and Vietnamese communities downstream are likely to incur the majority of negative impacts, such as delayed timing of the annual flood, demanding adaptation measures to mitigate introduced risks to current livelihoods. Rice cultivation in Cambodia and Vietnam are nominated examples.

The rice production system in the Mekong delta typifies the dependence of a highly efficient agricultural system on very specific, predictable timelines of natural resource occurrences. Minor variations in annual flood timings potentially jeopardise entire harvests. In contrast, many dam related discussions identify the benefits to downstream communities of the reduced risks of natural flooding as a function of increased storage capacity. In the case of the Mekong River, the benefits of impoundment related flood mitigation are not sufficient to compensate the costs of reduced sediment and nutrient loads associated with natural floods. The reduction of the positive side effects translates as increasing costs for fertilizers, erosion mitigation and the loss of agricultural land. Aligned with the reduction of natural flood risks is the risk of anthropogenic floods, likely to increase, as the independent operation of cascading dams will remain uncoordinated. Recent events have shown that dam operators do not always adhere to concession agreements and comply with agreed operational protocols. Non-compliance can create high amplitude flood peaks due to uncoordinated releases with detrimental implications for downstream communities, including Lao PDR communities.

Clearly, delayed floodwater, the loss of agricultural land and higher agricultural production costs constitute hydrological dynamics that are associated with negative implications for food security. The hydrological perspective defines a redistribution of benefits and risks from water, with negative impacts in Cambodia and Vietnam while Lao PDR and Thailand are predicted to experience a mix of positive (i.e. more water during the dry season) and negative consequences (i.e. loss of agricultural land).

Fish is another critical dimension of food security, which is largely affected by the introduction of fish passage obstructions. In particular, the negative impacts of hydropower on white fish are likely to reduce food security and further reduce already critical nutrition and protein deficiencies (Chaps. 3 and 5). Predicted losses of river gardens and non-timber forest products are likely to accelerate the decline of regional food security. The only counterweights to attenuate a devastating outcome for food security are the increased production associated with irrigation potential and dam fisheries. However, consistent with the hydrological perspective, these potential offsetting benefits of improved food security only occur in Lao PDR and potentially Thailand. Mekong communities in Cambodia and Vietnam are likely to bear negative consequences without the counter-balancing benefits of increased irrigation potential. The food security perspective indicates a redistribution of food security and potential benefit and wealth from Cambodia and Vietnam to Lao PDR and Thailand, consistent with the hydropower perspective.

The increased future demand of the energy sector provides the underlying imperative and rationale for increased dam development. Hydropower dam development is also the home to the largest economic incentives of this development strategy. The energy perspective identifies Thailand, China, Lao PDR and Vietnam as beneficiaries of improved energy security resulting from the increase in electricity generation. Additionally, Lao PDR would benefit from additional royalty and concession payments. Electricity generation and concession payments are aspects likely to interesting and unexpected flow-on effects. For instance, limited access to reliable and sufficient energy supplies is often identified as a constraint for further industrialisation and the ability to meet urban system demands. Removing energy constraints enables accelerated industrialisation and urbanisation, imposing further pressure in food security. However, combining energy generation with well-designed development strategies, could improve household employment opportunities in secondary and tertiary sectors.

The generated revenue harbours the potential to provide new secondary and tertiary livelihoods in Lao PDR and, thereby, not only reducing poverty but also improving food security. If cheaper food is produced in neighbouring parts of the wider Mekong Region and income could be substantially increased for larger parts of the Lao PDR population food security would improve. However, these development strategies have not been explicitly articulated, which contributes to largely negative and apparently biased assessments of the impact of hydropower on food security. Once identified, specific development strategies that utilise the revenue of the energy sector could be included in assessments and are likely to contribute to a much more balanced view.

Similarly, the provision of electricity to the emerging manufacturing and service sectors of Thailand, China and Vietnam are likely to generate more employment opportunities. While fuelling urbanisation processes and many aligned problems the income effects are likely to be positive, also improving food security in these three countries.

However, these positive processes are long-term eventualities, while the negative impacts on communities are immediate. The insight of temporal dis-concordance also points to distributional and economic differences dams can have on different parts of society. The effects discussed in Chaps. (2, 3, 4, 5, 6 and 7 are largely negative at the household level while the positive implications tend to be redistributed and confined to a select cohort of individuals; that is a concentration and conglomeration of wealth potential due to shifts in resource access.

Societal wealth redistributions combined with the substantial differences between the five countries suggests increasing pressure on households in Cambodia, Vietnam and Lao PDR. Bouapao (Chap. 5) argues that affected households might migrate into more promising areas, such as Thailand.

Migration is likely to redistribute not only population but also poverty and the need to provide attendant and necessary services. Previous Chapters have largely portrayed migration as a negative process, despite multiple associated benefits, such as the provision of labour, new livelihoods and cultural exchange. The assumption of large scale migration may have been an important factor in the negative evaluations. Large scale migration correlates with social and political conflict, which in turn harbours repercussions for economic growth and poverty levels. Economic growth remains a central pillar of national policy objectives articulated by all six countries in the wider Mekong Region. Thereby, risking food security and redistributing water access becomes a matter of sovereign concern for all six countries as political stability and economic growth are at risk.

3 Water Diversion

The dominant hydrological implication of water diversions is the reduction in dry season flow as diversions are geared to meet dry season irrigation demands (Chap. 2). The potential for increased irrigation water could improve food security, although current plans in areas of northeast Thailand, where water diversion is mostly discussed, target increased production of energy and biomass crops. Increased irrigation often triggers decreasing water quality due to the increased application of agro-chemicals, introducing contaminate hot spot for communities in Thailand and Lao PDR. From an energy perspective water diversion can increase the availability of irrigation dependent biofuels, which reduces –similar to the predicted consequences of hydropower – the regions dependency on fossil fuels.

The predicted array of water-food-energy nexus characteristics differs for water diversions compared to those predicted for hydropower. Diversions have largely localised effects with some negative effects downstream that become more pronounced as the volume of diverted water increases. However, the investment shifts benefits from downstream communities to communities with access to the irrigation scheme. Positive dynamics of water diversions could either coincide with positive energy or positive food security, depending on the class of irrigated crops. Downstream communities would face the opposite result in nexus dynamics. The localised changes in benefit distribution introduce incentives at the household level to migrate as livelihood opportunities improve elsewhere and deteriorate downstream.

4 Sea-Level Rise

Sea-level rise and potential response measures do not constitute development strate gies. We decided to include them because sea-level rise is an important process and part of the reality of decisions in the wider Mekong Region. Therefore, careful consideration of the consequences of sea-level rise and potential adaptation measures can make a substantial difference when assessing the cumulative impacts of other development decisions. For instance, hydropower can have the potential to mitigate salinity intrusion due to sea-level rise and, thereby, creating positive effects for Vietnamese communities in the Mekong Delta.

Sea-level rise is projected to affect all delta floodplains in the wider Mekong Region. Community and livelihood vulnerability is a function of floodplain size, elevations closer to sea levels, population size and the degree of land utilisation. Most attention has been focused on Vietnam's Mekong delta because of the combination of population, utilisation, size and elevation. Most available data and analysis and, therefore large parts of the relevant discussion on sea-level rise in

Chaps. (2, 3, 4, 5, 6 and 7 are focused on Vietnam's Mekong delta. However, sea level rise is likely to affect both rural and urban areas located in coastal regions. The consequences of sea-level rise throughout the coastal region in the wider Mekong Region may be highly underestimated, as many urban centres are located in floodplains.

Increasing water salinity levels that affects both agricultural and urban water quality and access has been a primary focus of hydrological analysis. Rising sea levels are likely to alter the balance of freshwater and saltwater in coastal aquifers through increased salt water intrusion, reducing potable drinking supplies and increasing the vulnerability of urban groundwater-based systems. Consequently, urban centres in particular could face substantial financial challenges.

From the perspective of food security it is essential that households adapt, for instance shifting the primary focus of agricultural production from rice into aquaculture. If biophysical conditions reduce land use options and livelihood diversity declines, communities in larger areas are often confronted with higher levels of vulnerability due to economic fluctuations or diseases. The management of such vulnerabilities can come at high public costs, while unmanaged situations can lead to even higher costs in situations of crisis.

The energy sector might experience higher demand if aquaculture increases the household income in far more communities. In urban areas water treatment investments are likely to increase, as drinking water is likely to require additional processing. Additionally, existing infrastructure of the energy sector in low-lying areas could require upgrading due to sea-level rise. All the remediation investments are associated with attendant increases in national GDP.

The consequences for communities is highly uncertain; opportunities created by aquaculture could improve livelihoods while natural and economic uncertainties are likely to impose more risks and threats to human life. Ultimately, a tipping point could be crossed triggering sudden out-migration from low-lying areas. However, people live in many parts of the world with high risks, which makes it unlikely that a sudden exodus due to sea-level rise would occur over the next 50 years.

Strategies in response to sea-level rise in Vietnam's Mekong Delta are largely discussed as a binary decision; either to build large-scale dykes and sluice gates or to invest in land use change, including shrimp, salt resistant rice varieties and other crops. The effectiveness of these two strategies, dykes or land use change, depends on the characteristics of upstream changes. For instance, hydropower operations and dam releases have the potential to combat large-scale salinity intrusion. However, reservoirs are not managed for this purpose but are largely operated in response to rainfall and the financial incentives to supply the peak demand of electricity markets. A revised portfolio of economic mechanisms and the institutional coordination of upstream dam operations would be required to effectively reduce salinity levels. As it seems unlikely that Vietnam would make compensation payments to reservoir operators and that upstream dam releases will be coordinated, reduced salt water intrusion due to hydropower operations is consigned to a residual side effect. The predicted levels of dry-season flow increases would not make dykes or other adaptation measures obsolete. On the contrary, some research predicts that the combination of hydropower (mainstream and tributaries) and water diversion is likely to increase salinity intrusion (Nhan et al. 2012; Toan 2012).

Dykes and sluice gates are energy intensive during construction, and although boat passage might be limited, it does not seem likely that food security would be reduced. Conversely, if effective, dykes could maintain rice production and safeguard conditions for agricultural progress leading to increased rice production.

Initiatives concerned with land use change would seek to maintain or improve food production. Some research predicts that without dykes and with upstream developments of reservoirs and water diversion, food production could be maintained or even increased by using more salt-resistant rice varieties and cultivating more shrimp in coastal communities (Nhan et al. 2012). However, the highly efficient rice production system of Vietnam's Mekong Delta depends on stable wet season flows and substantial sediment and nutrient transports. A singular reliance on salt-resistant varieties may not alter the exposure levels and vulnerability of the rice sector to upstream hydropower and water diversions. Similarly, shrimp depends on stable salinity levels, which might not be guaranteed in dry years, when reservoirs need longer periods to replenish and more water is diverted.

5 Rubber Plantations

The increase in industrial plantations of rubber in the wider Mekong Region introduces risks for a growing number of communities as dependency on the rubber and latex market increases. The hydrological implication still lacks a robust study, in particular if whole provinces undergo land use conversion from diverse land cover into a monoculture, water-intensive crop. Changes in the local hydrology and in the microclimate are likely potential consequences, reducing dry-season flows for downstream communities.

The ramifications of industrial rubber production on local food security appear to be more pronounced when compared to water security. Additional land allocated to rubber production results in income from cash crops replacing local food production. Changing land use increases the dependency on food-producing areas elsewhere. When connected with income vulnerability, aggregate food security potentially declines for large areas in the Mekong Region.

The energy sector is unlikely to experience large impacts from increased land allocated to rubber production, apart from a decrease in biofuels and energy crops. Depending on the energy requirements of rubber processing, energy demand might increase.

As long as the rubber markets remain robust, typified by high latex prices, in-migration is likely to increase in rubber producing areas. A prolonged decrease in rubber prices has the potential to catalyse a household income crisis, typical of an extensive, capital intensive and long maturation monoculture land use system, which translates into substantial out-migration.

6 Mining

Lazarus (Chap. 7) explained the relevance of bauxite mining for the wider Mekong Region. Hydrological implications emerge as a consequences of the energy demands of bauxite processing, as hydropower is a likely and substantial component of an electricity generation solution. Extensive impoundments for hydropower development change the hydrograph of Mekong tributaries with attendant consequences for downstream communities, i.e. late onset of annual floods. Additionally, large-scale mining operations such as bauxite can change localised geo-morphological characteristics and, therefore, act as a water diversion between two watersheds.

The loss of cropland and non-timber forest products for many years will introduce additional pressure on food security. However, if income levels are improved for the local populations residing in mining affected regions, food security might be safeguarded.

Energy demand is likely to increase due to the mining and processing activities, adding further pressure to install more power plants, such as hydropower. With more hydropower food access might experience further decline. Bauxite processing also demands large electricity base loads, usually reliant on fossil fuel generation, which requires large volumes of cooling water. Both options are likely to alter local hydrographs.

With new employment opportunities people are likely to migrate into typical mining towns, an experience repeated many times around the world.

7 Transportation Infrastructure

In comparison to the changes discussed above, additional transportation infrastructure does not coincide with substantial weakening of the water-food-energy nexus. This Volume assumed an extended railway network, which could have negative effects on food security if land access rights shifted significantly. Energy demand is likely to increase although, from a regional perspective other, more energy intense transportation modes could be replaced. Improved transportation options could improve market access and reduce migration impediments, which mean that connectivity in the wider Mekong Region will further increase.

8 Cumulative Cross-Sectoral Assessment

Typically, development strategies are assessed individually and independently. Such artificial separation ignores the fact that many investments impact on each other. Some investments can offset each other in regards to particular indicators such as livelihoods or food security. Typically, if a development decision has negative effects on an important sector but positive impacts on another sector, one corrective strategy might contemplate an additional investment to offset the negative without compromising the positive effects. Similarly, many investments, if combined, could achieve more than the sum of individual strategies. However, unintended side effects can also have synergistic effects.

The combined impact of all six development investments discussed in this Volume does not clearly benefit one singular dimension of the water-food-energy nexus. In contrast, a mosaic of mixed trade-offs emerges with developmental pressures shifting from one sector and country to another. However, the magnitude of the impacts caused by mainstream dams dominate with the assessment appraised from multiple nexus perspectives, at least for the lower Mekong Basin, and possibly beyond due to migration dynamics. Clearly, none of the other five development strategies are predicted to have the potential to mitigate the hydrological implications of mainstream dams. In contrast, water diversion, rubber plantations and mining are likely to exacerbate the hydrological implications for downstream communities. Food security, on the other hand, has a few positive aspects as food access could improve with increasing household income, provided by more irrigation, aquaculture, and employment on plantations and mines. However, this particular nexus ensemble assumes stable food prices, which in the global economy remains highly uncertain. Most strategies that have been discussed are likely to increase energy demand, while hydropower is unlikely to provide a substantial proportion of the energy portfolio projected for the wider Mekong Region.

It seems that most evidence points to a wider Mekong Region development trajectory that is characterised by increasing vulnerabilities due to encroaching monocultures, rising sea levels, dependency on purchased rather than self produced food and paid labour. A multi-factor transformation of the wider Mekong Region introduces criticalities that require close monitoring and policies that effectively balance risks within the nexus. However, effective governance mechanisms are likely to incur high economic costs. With larger areas under monoculture (i.e. rubber and shrimp) and the loss of many other food sources (i.e. fish and riverside gardens) the wider Mekong Region is likely to experience economic difficulties that will translate into food access problems. If national governments or a new regional governance mechanism seeks to establish safeguards for a food crisis, costly logistical measures are required for a region that thrived for many centuries with relative food abundance.

Contingent on the identified food security risks the reallocation of water resources from downstream to upstream communities is a critical factor, albeit subject to different time scales. The combined impact of the six development investments we discussed could have positive impacts on communities in Lao PDR and Thailand. These income effects may even occur in the mid-term. However downstream communities are likely to face short-term external costs and negative consequences. If adaptation options for downstream communities are not developed and implemented in the short-term, the risks associated with upstream developments might initiate a crisis in downstream countries. Such a scenario would challenge the current period of political stability characteristic of the wider Mekong Region. Clearly, whilst a possible future scenario, the possibility remains that most identified risks increase at a regional scale, which if recognised, introduce additional challenges for national policies and initiatives.

Avoiding worst-case scenarios requires effective mechanisms to avoid triggering unintended ripple effects. Annex A-F illustrates some of these highly complex ripple effects for the six development strategies assessed in this Volume. Connectivity is very complex, characterised by multiple interactions, both reinforcing and opposing, and emergent ripple effects. Emergent ripple effects are generally understood to have negative consequences. However, the obverse also applies: ripple effects can be understood as outcomes that can be achieved with minimal additional effort and resource investment. Benefiting from the potential for positive ripple effects requires a thorough understanding of the connectivity of the system.

The cross-sectoral perspective of the trans-boundary context connects many possible ripple effects. Importantly, a few emergent system nodes highlight criticalities that require careful management independent of the actual development strategies.

Critical nodes represent a constellation of factors, that combined, influence the degree of nexus linkages and as a corollary, livelihood outcomes and poverty levels. The magnitude and direction of poverty reduction and livelihood levels are important emergent outcomes of node dynamics. High levels of uncertainty surrounding node dynamics and factor interaction translate as unreliable assessments of potential poverty reductions. The assessments discussed in this Volume indicate that poor understanding of node interactions is likely to result in reduced effectiveness to alleviate existing poverty levels.

Directing efforts to improve the understanding of node dynamics and developing instruments geared to managing interacting factors represents an alternate and potentially more effective policy strategy than a singular focus on poverty itself.

Critical system nodes identified as part of the expert panel assessment are:

- Fish stocks are largely discussed within the context of mainstream dams. However, it emerges as a critical resource that requires careful management and new governance solutions independent from mainstream dams.
- Impacts of land use change and irrigation are highly uncertain. If land use change and irrigation are managed primarily as an instrument to alleviate poverty the development of the wider Mekong Region is likely to experience a more sustainable development. Risks in the nexus need to be managed carefully with particular attention to monoculture dependencies.
- Migration dynamics depend on a portfolio of available livelihood options and determine a critical factor influencing political stability.
- Labour shifts from primary to secondary sector employment, in particular as part of an underpinning urbanisation trend. Investing in the necessary urban infrastructure and in maintaining rural livelihood options will be a key factor in safeguarding social and economic stability.
- It will be critical to effectively manage energy demand (as opposed to the singular management of power supply).

These aspects emerge from the system diagrams as critical nodes. Clearly, none of the six large-scale development strategies discussed in this Volume contributes to these criticalities without creating associated trade-offs. Instead, the combined effects increase risks in all of these dimensions, weakening the regional water-food-energy nexus. The set of critical nodes allows the derivation of alternative development investments and trajectories that could improve and strengthen the water-food-energy nexus in the Mekong Region.

- The development of a regional monitoring and governance system for fishing and fish migration, with effective enforcement mechanisms, could substantially contribute to the strengthening of the nexus.
- Incentives can be designed to minimise the extensive development of monocultures and to enhance diverse land uses. This requires investments in improved productivity of small-scale landholders and transboundary mechanisms for guiding foreign direct investments, in particular to protect existing land titles and tenure.
- Lower uncertainty in the food domain coupled with improved returns, diminishes the risk of large-scale migration.
- With more stable incomes in rural communities, urban influx is likely to decrease reducing the growth rate of future energy demand.
- Additional investments in promoting energy saving through a raft of initiatives including building and construction ordinances and consumer choices could strengthen the energy domain without compromising water access and food security.

The suggested bundle of strategies could shift the system on a pathway to sustainable development of the wider Mekong Region. Water diversion contributes to this articulated vision if food security is not compromised by irrigating largely energy crops and if kept at a scale that does not affect downstream communities. Mining contributes to a sustainable development trajectory if local communities experience a clear benefit and not a net loss across the portfolio of contextual livelihoods. Additionally, implications for downstream communities would need to be managed carefully to avoid unsustainable effects. Transportation infrastructure could improve the access to markets and strengthen land use diversification.

In contrast, mainstream dams and large-scale rubber plantations emerge as strategies that increase nexus-related risks to a point where regional political stability might be compromised. Sea-level rise is likely to be a trend that could amplify similar destabilising effects in the long-term if critical thresholds are exceeded without adequate adaptation measures put in place.

In summary, the six development strategies assessed in this Volume are largely driven by short-term profitability instead of motivations of long-term stability and sustainable development. Lessons made in developed countries are instructive in identifying the long-term costs associated with short-term optimisation strategies and the mutualism of economic gains and the political stability of a larger region.

9 Reflections on the Expert Panel Process

Most investments that occur to meet the objectives of one particular sector have side effects on other sectors. In a highly connected region such as the wider Mekong Region, emergent side effects are likely to transgress national and natural boundaries and borders. Comprehensive and coherent assessments of investments require reflection on these cross-sectoral and transboundary implications. The Exploring

Mekong Region Futures expert panel was formed to conduct such an integrated assessment based on existing research combined with expert opinion.

The advantage of an expert panel is that tractable results can be generated relatively quickly as additional field work or modelling is not required. Once funding has been secured, the required expertise identified and commissioned, the sectoral assessment can be available within a few months.

Presenting sectoral assessments and tracing cross-sectoral impacts in a workshop was very effective. However, our aim was to inculcate a process that assisted participants to reflect on subjective biases. Therefore, we did not disclose the initial causal factors when we asked sectoral experts to consider the primary impacts identified for other sectors. A similar rationale applied to invitations to identify secondary and tertiary effects. Brainstorming such ripple effects was difficult because sector experts repeatedly insisted on identification of the initial causal factors, claiming secondary effect dependency. We contend that these claims represent evidence of strong biases in the research community and actual limitations in delivering an unbiased assessment. Clearly, not all assessments can be balanced as in many situations investments may have a positive effect in one sector and negative (side) effects on other sectors. Thus, assessments can appear unbalanced in single sector evaluations, as the proposed effects have a singular focus and are not balanced within each sector.

Chapters 2, 3, 4, 5, 6 and 7 combine effects, personal biases and the fact that influences individual sectors can be naturally uni-directional. We believe that our approach accounted for biases, albeit in an uncontrolled manner, and allowed for a more objective tracing of ripple effects expressed in the system diagrams (see the Annex). Inviting additional experts to this workshop process was another effective way to capture diverse opinions. If repeated, we would be inclined to invite even more additional experts to the workshop process to extend the robustness of the results. Further reducing biases would have compromised expert opinions and introduced an unwarranted level of censorship.

Chapters 2, 3, 4, 5, 6 and 7 represent views that dominate the research and consultancy contributions of practitioners active in the wider Mekong Region. The key insight we gained is that a large cohort of potential benefits remain largely unspecified, due to the lack of details for some of the development strategies as opposed to extant biases. We identified three aspects that could have large-scale positive impacts:

- The accrued hydropower royalties in Lao PDR could be assigned to effectively create employment opportunities, improve existing livelihoods and, thereby, reduce poverty. This could also have positive regional implications. However, details of actual or potential royalty distribution are not publicly available. This limits the ability of such assessments to provide a balanced view.
- The provision of large-scale transportation has the potential to improve market access for many and, thereby, generate additional income. However, the lack of detail of such investments and the lack of follow-up strategy does not provide a clear picture of potential benefits, leaving the evaluation of benefits to speculation rather than evidence based.
- Sea-level rise threatens to impose substantial potential costs. Being perceived as a long-term trend, focussed attention is confined to a very few areas, in particular

in Vietnam's Mekong Delta. The omission of other low-lying urban and rural areas from research and policy scrutiny increases their exposure to unnecessary risks. Including sea-level rise effectively in this integrated assessment would have required much more information on potential damaged for all coastal communities in the wider Mekong Region and details about adaptation options.

Identifying these three points required a careful examination and synthesis of the expert panel assessment. We propose that the expert panel process would benefit from additional iteration of deliberation, detailing these three aspects, conducted in association with decision makers to achieve a more comprehensive, less speculative and balanced assessment.

However, assessing development strategies in environments as complex as the wider Mekong Region is difficult and cannot be absolutely balanced, unbiased and comprehensive. A multitude of diverse and competing claims and values nurture the normative dimension of discussions in the wider Mekong Region, where value-based debates and contradicting worldviews are the norm, and we contend should be promoted. This Volume aims to capture some of these views and contribute to a better understanding of the cumulative cross-sectoral and transboundary effects of multiple development decisions. Independent of the normative challenge, a considerable disadvantage of an expert panel approach is the variable ability of panels to consider complexity, an insight consistent with the nuanced claims of Delphi Technique proponents.

Accounting for the variable individual and collective comprehension of the multitude of potential dynamic interactions represents a formidable challenge to the expert panel approach. The system diagrams illustrated in Annex A to F captures only the most critical connections for each development investment. Overlaying all diagrams would provide the reader with a much more comprehensive visualisation of the most important connections in a cumulative perspective. Unfortunately, the resulting diagram would be illegible and, therefore, without any benefit. This stresses the cognitive limitations humans face when processing complexity. Experimental research has already accumulated substantial evidence of human cognitive limitations. Therefore, expert panels can be designed to step sequentially through sectoral and cross-sectoral impacts, revealing a rudimentary but highly insightful view of complexity. However, providing a reliable quantitative assessment that considers the majority of relevant connections is a challenge the expert panel method is unlikely to overcome. Computational methods are increasingly deployed as a research tactic to increase the capacity to understand higher levels of complexity. The expert panel approach conducted for the wider Mekong Region has been complemented by the development of a complex systems model and a series of other research methods, see Smajgl et al. (in review).

In summary, expert panels can provide timely assessments but can be biased and face challenges in highly complex situations. Conducting further iterations with decision makers relevant to the investment proposal and introducing complementary methods could alleviate some of the challenges but would demand considerably more time. Unfortunately, many development strategies are implemented without the necessary time for adequate assessments, which leaves many outcomes to chance and increases the risk of unsustainable development pathways.

References

Nhan, D.K., V.A. Phap, T.H. Phuc, and N.H. Trung. 2012. Rice production response and technological measures to adapt to salinity intrusion in the coastal Mekong delta. A report to the exploring Mekong region futures project; CSIRO Climate Change Adaptation Flagship, Canberra Australia. http://www.csiro.au/science/MekongFutures

Smajgl, A., J. Ward, T. Foran, J. Dore, and S. Larson. in review. Visions, beliefs and transformation: Methods for understanding cross-scale and trans-boundary dynamics in the wider Mekong region. *Global Environmental Change*.

Smajgl, A., and J. Ward. in review. A design protocol for research impact evaluation: Development investments of the Mekong region. *Research Evaluation*.

Toan, T.Q., N.H. Trung, and D.K. Nhan. 2012. The Mekong future project: Draft report on the hydrological simulation. A report to the exploring Mekong region futures project; CSIRO Climate Change Adaptation Flagship, Canberra Australia. http://www.csiro.au/science/MekongFutures

Annex A: System Diagram Illustrating the Connectivity Arising from Hydropower Development

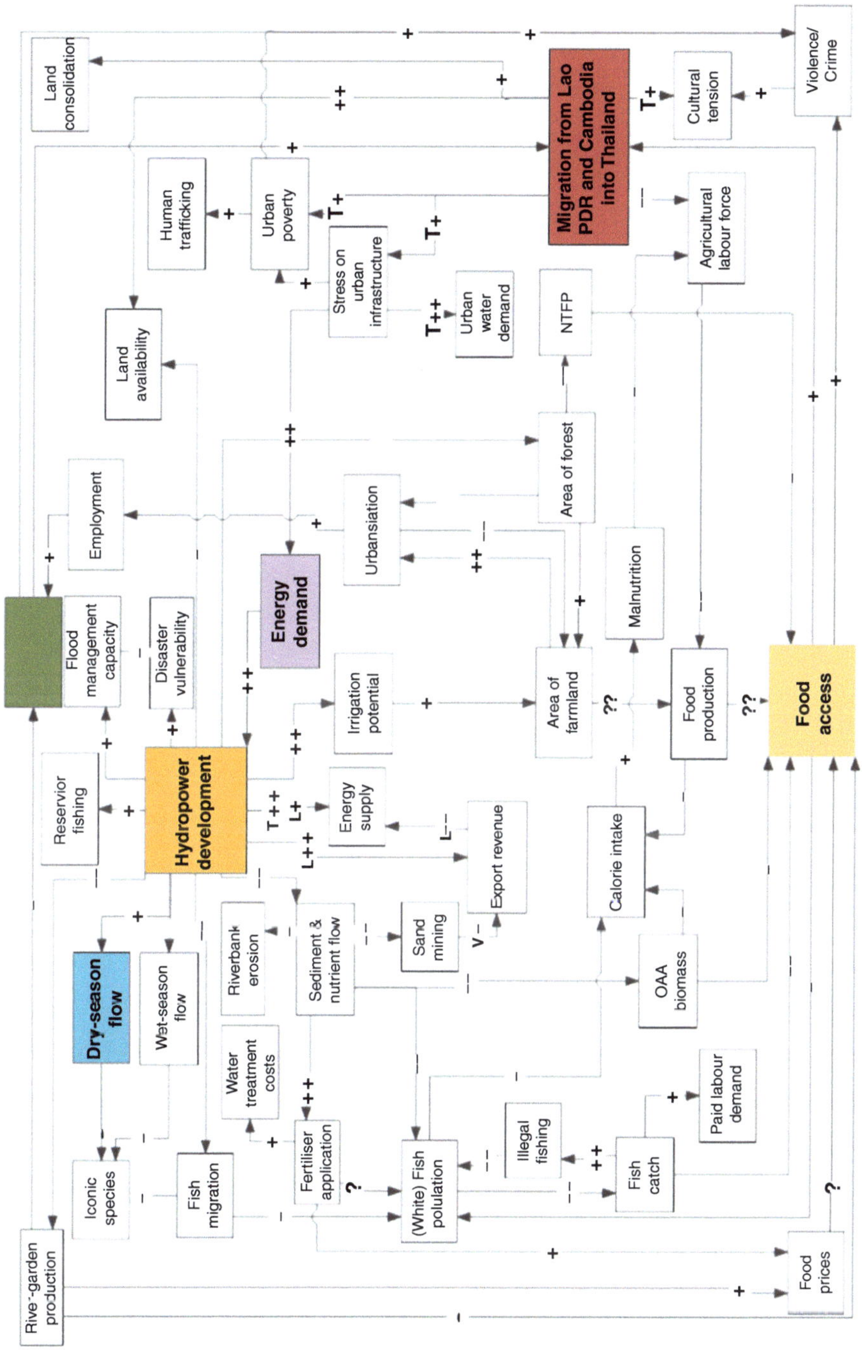

A. Smajgl and J. Ward, *The Water-Food-Energy Nexus in the Mekong Region: Assessing Development Strategies Considering Cross-Sectoral and Transboundary Impacts*, DOI 10.1007/978-1-4614-6120-3, © Springer Science+Business Media New York 2013

Annex B: System Diagram Illustrating the Connectivity Arising from Water Diversions

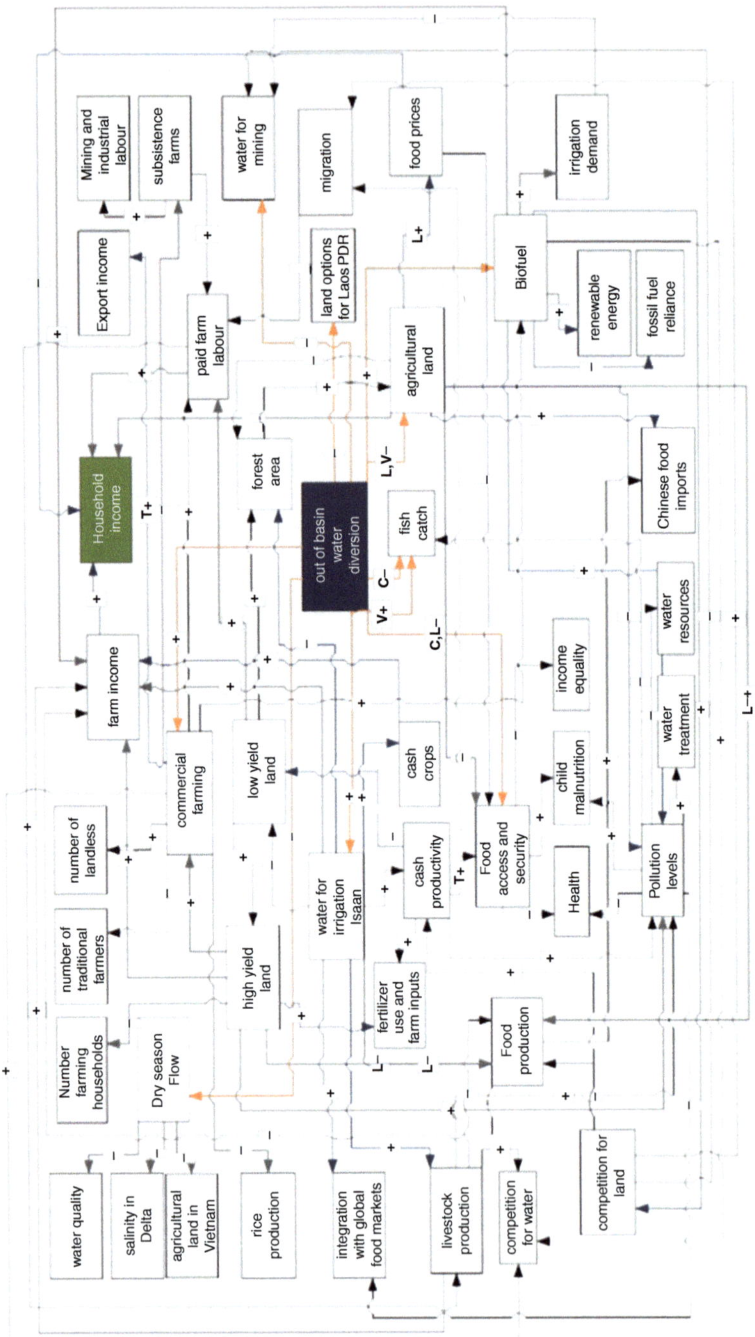

Annex C: System Diagram Illustrating the Connectivity Arising from Industrial Rubber Plantations

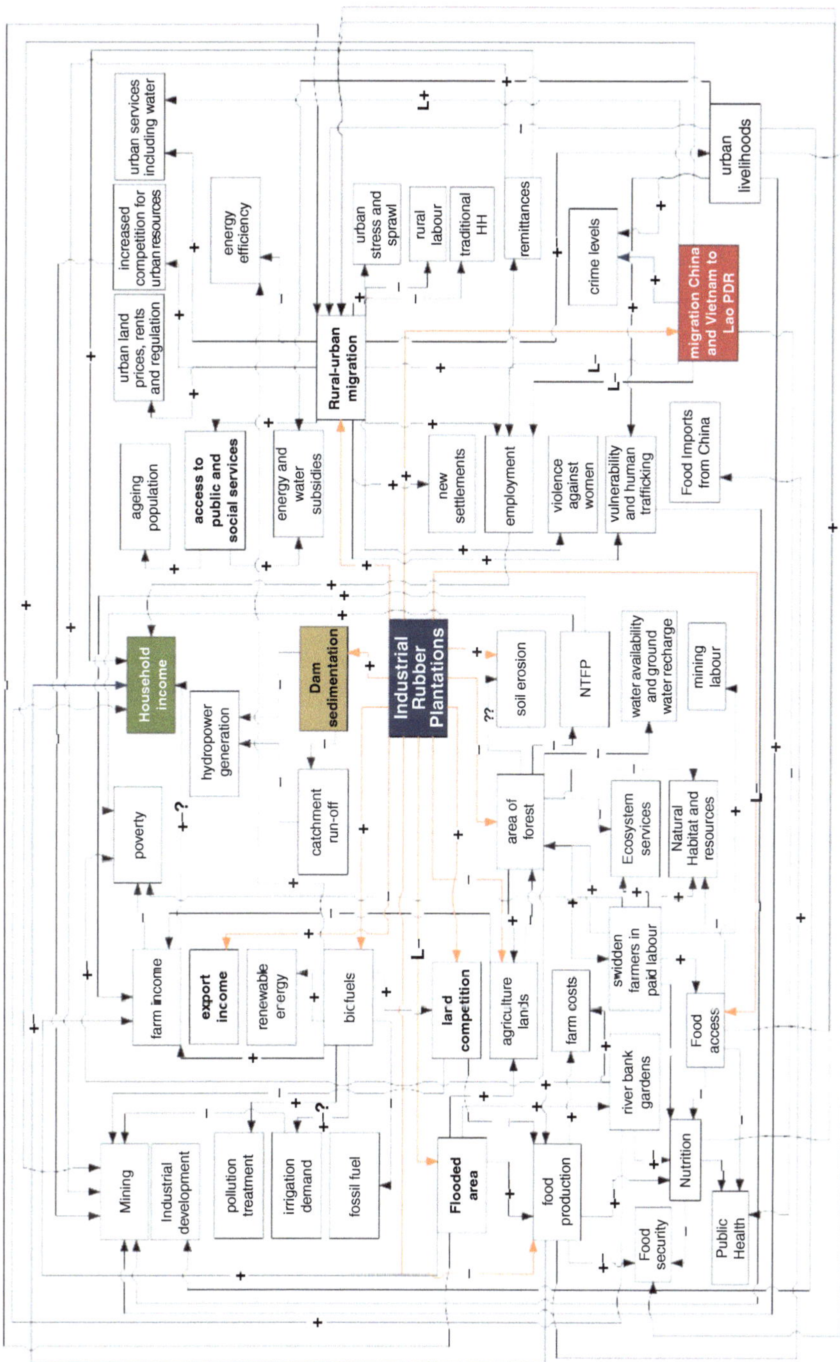

Annex D: System Diagram Illustrating the Connectivity Arising from Sea Level Rise

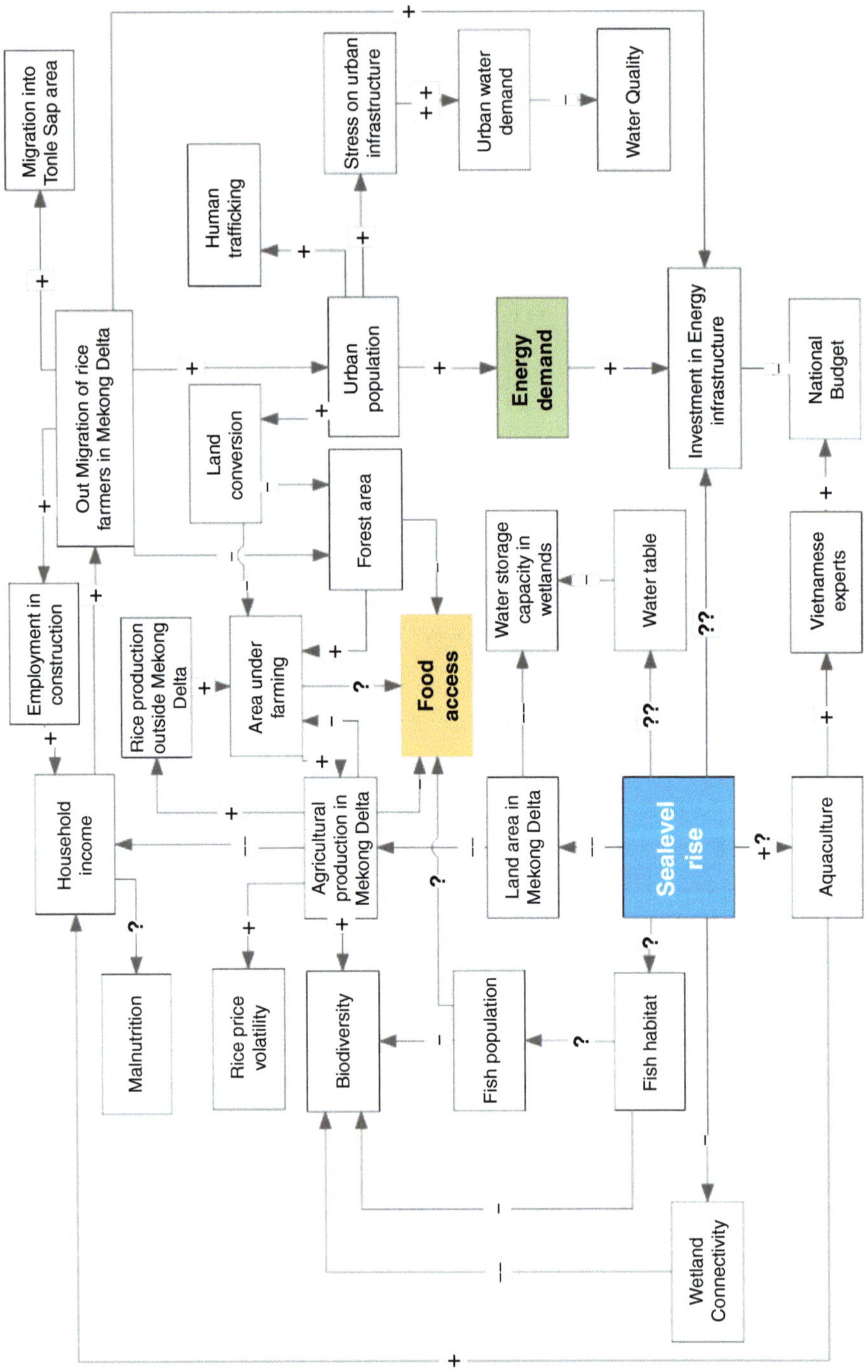

Annex E: System Diagram Illustrating the Connectivity Arising from the Kunming to Phnom Penh Railway Connection

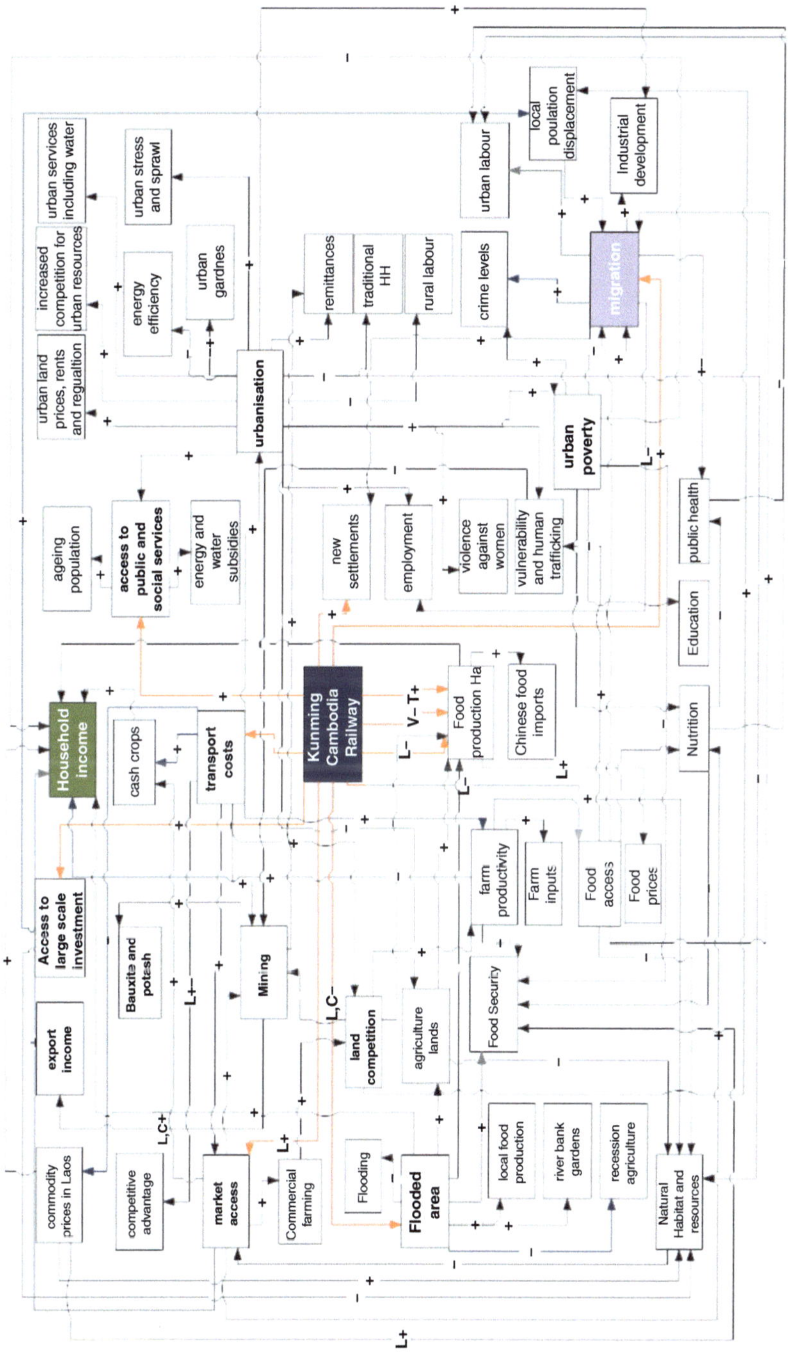

Annex F: System Diagram Illustrating the Connectivity Arising from Bauxite Mining

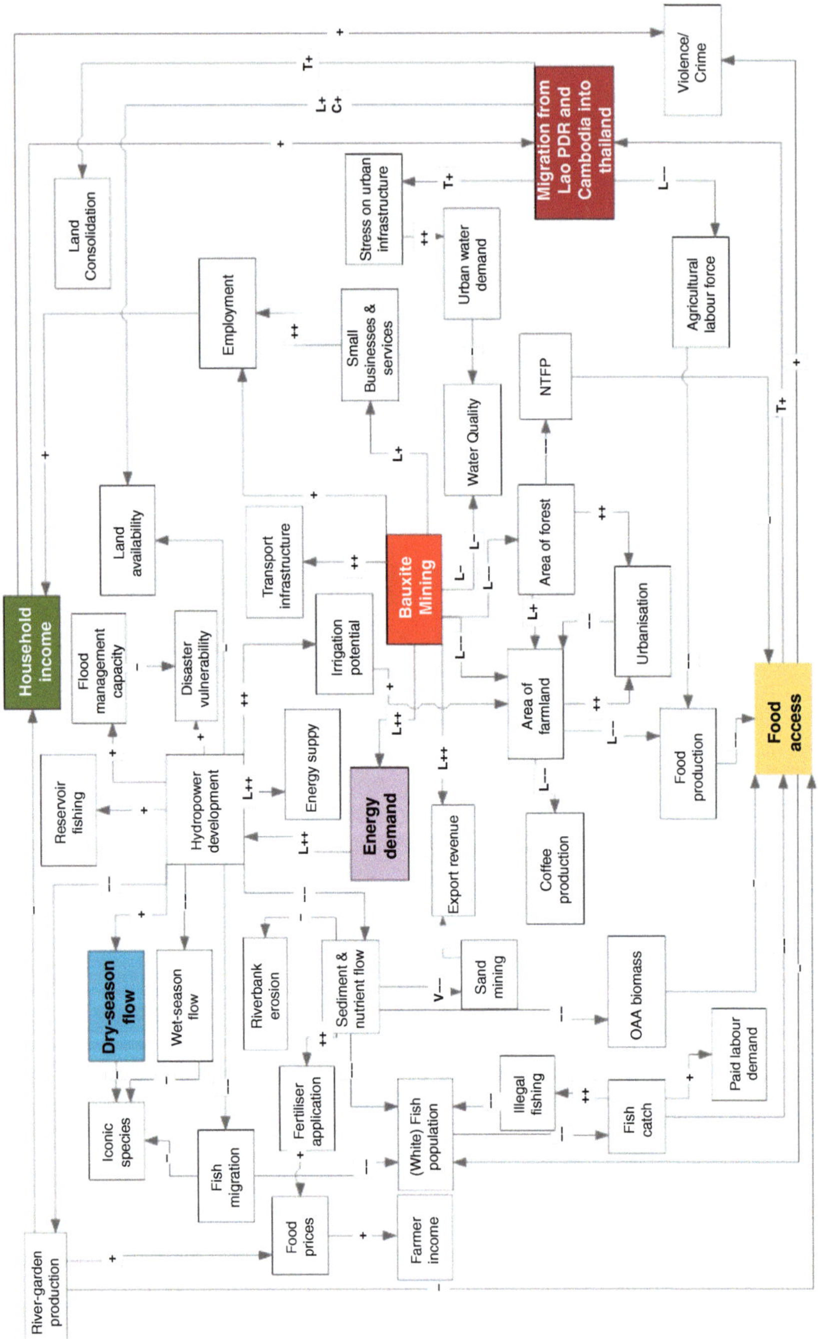

Author Biographies

Lilao Bouapao has recently completed his 6-year assignment as a Senior Social Science Specialist at the Mekong River Commission (MRC) Secretariat. He is currently working as a Consultant for a number of organizations, including MRC, Mekong Development Center (MDC), the Challenge Program on Water and Food (CPWF), and International Finance Corporation (IFC) of the World Bank. His focus has been on statistics and impact assessments, including undertaking censuses, surveys and case studies at both national and regional levels, specializing in social sciences, water resources science, and rural development.

Tira Foran, a specialist in governance of hydropower and electricity planning in mainland southeast Asia, studies social change around energy and resource use. Dr. Tira Foran joined CSIRO in June 2010. His recent work has focused on transitioning to more sustainable energy systems, particularly how to improve the governance of water and energy resources using techniques such as dialogue, participatory scenario building and modelling. Dr. Foran is currently involved in new research around urban sustainability and sustainable lifestyles. Dr. Foran's background is in contentious politics and natural resource management. He has over 10 years of experience in environmental policy analysis, working in both the private sector and academia. His past work has been with Chiang Mai University, Thailand; International Union for Conservation of Nature (IUCN); The World Conservation Union, Thailand; and Environmental Defense Fund, California, USA.

David Fullbrook is a sustainability analyst. He has provided trans-disciplinary perspectives on agribusiness, common-pool resources, energy, food security, hydropower, market assessment and project planning to public and private clients with interests in China and the Greater Mekong. David received an MSc Asian Politics from the University of London's School of Oriental and African Studies and an MSc Environmental Management from the National University of Singapore. During 2012-2013 he is developing his expertise in sustainability, particularly ecosystem services and the consequences of climate change especially for sea levels and cities, while studying for an MSc Ecological Economics at the University of Edinburgh.

A. Smajgl and J. Ward, *The Water-Food-Energy Nexus in the Mekong Region: Assessing Development Strategies Considering Cross-Sectoral and Transboundary Impacts*, DOI 10.1007/978-1-4614-6120-3, © Springer Science+Business Media New York 2013

Kate Lazarus has worked in the Mekong Region for 12 years on water resources management, hydropower and environmental governance issues. She has researched policy issues such as the role of China in the Mekong Region, environmental flows, nexus of rights and environment, stakeholder engagement and hydropower decision-making. She has worked for such organisations as Oxfam America, The International Union for Conservation of Nature (IUCN), the M-POWER (Mekong Program on Water, Environment and Resilience), and the Challenge Program on Water and Food (CPWF).

Mr. Xing Lu, Associate Professor at the Institute of Southeast Asian Studies, Yunnan University. He is interested in environmental politics and has been studying governance issues of natural resources, cross-border trade and investment in the wider Mekong Region.

Sokhem Pech (Ph.D., (HON) M.A., L.L.M.) is a senior environmental governance specialist, Hatfield Consultants Partnership, North Vancouver, BC. Canada. Pech is also a Co-chair, Steering Committee, Mekong Program on Water, Environment and Resilience (M-POWER). Sokhem Pech has over 28 years of experience working at a senior management level with national governments, inter-governmental organizations, international multi-disciplinary research consortiums, and consulting companies in Asia and North America. He specializes in institutional capacity assessment and development, strategic planning and program development, Climate Change adaptation planning, monitoring and evaluation, policy interface and communication, sustainability assessment frameworks, environmental and climate related risk management, and sustainable policy development.

Alex Smajgl works on developing participatory processes and integrated methodology in order to assess triple bottom line outcomes. Geographically, his work is focused on the Mekong region, Indonesia and the Great Barrier Reef region. In order to address stakeholders on various levels his participatory work uses a multi-scale assessment approach. Dr. Smajgl includes grass-root level methodology like agent-based modelling with the aim of simulating disaggregated system behaviour. In order to address stakeholders on various levels Dr. Smajgl's quantitative modelling work uses a multiscale assessment approach. Questions regarding how social ecological systems self-organise are crucial for his work, especially how social systems create and change rules to govern natural resource use. One of Dr. Smajgl's recent projects facilitated a discussion of potential impacts of energy policy options on poverty and deforestation in Indonesia. This integrated approach involved three levels of Government. In his current work Dr. Smajgl leads a participatory approach that looks into water-food-energy decisions in the wider Mekong region, likely trade-offs and transboundary dynamics. This work is conducted in partnership with many government and research agencies in Cambodia, China, Laos, Thailand and Vietnam. Dr. Smajgl joined CSIRO Sustainable Ecosystems (now Ecosystem Sciences) in 2003. His previous research involved quantitative sustainability assessment of post-Kyoto policies, and analysing and simulating impacts of climate policy instruments on natural resource issues, global and national energy aspects, and international trade.

John Ward (B.Sc. (Hons), Ph.D.) is a natural resource and behavioural economist whose research focuses on the ex ante testing of institutional settings and policies to manage natural, common pool resources. Dr. Ward's current research explores likely responses to water governance architectures, the cost effectiveness of hybrid policy instruments, and evaluating the alignment of incentives with community value orientations. He is a senior researcher with the Environmental Development Group of CSIRO Ecosystem Sciences, currently deployed in the Mekong region researching the nexus of food, water and energy security. Dr. Ward has a background in the biological sciences, natural resource and institutional economics, the forest industry and academia.

MIX
Papier aus verantwortungsvollen Quellen
Paper from responsible sources
FSC® C105338

If you have any concerns about our products,
you can contact us on
ProductSafety@springernature.com

In case Publisher is established outside the EU,
the EU authorized representative is:
Springer Nature Customer Service Center GmbH
Europaplatz 3, 69115 Heidelberg, Germany

Printed by Libri Plureos GmbH
in Hamburg, Germany